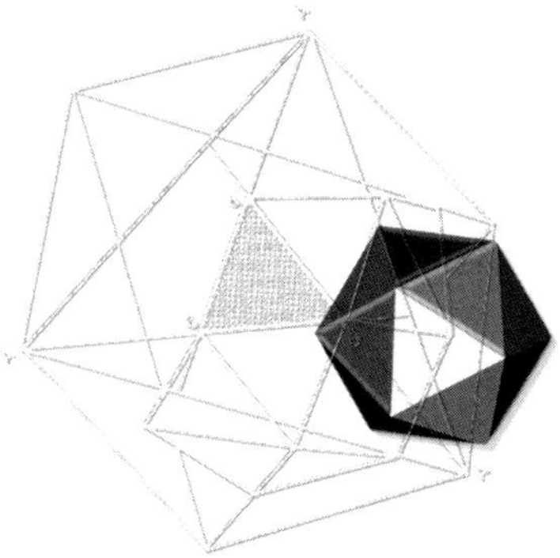

An Anthology of **Structural Morphology**

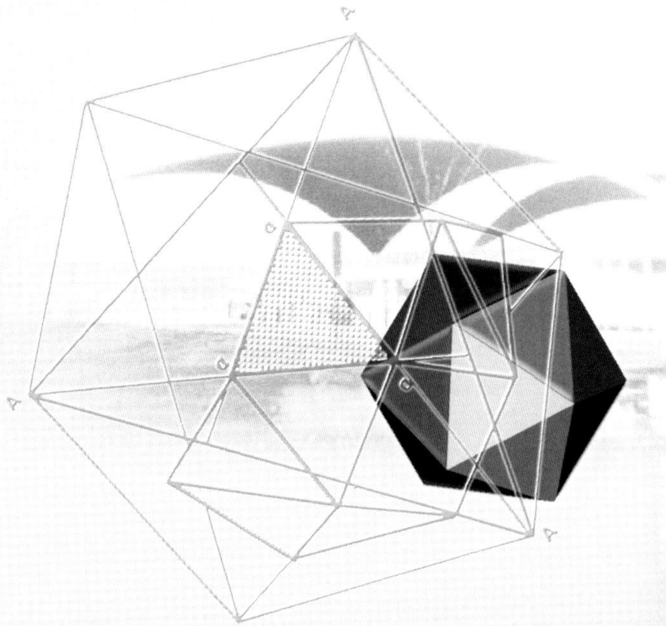

An Anthology of Structural Morphology

edited by

René Motro

Université Montpellier 2, France

World Scientific

NEW JERSEY • LONDON • SINGAPORE • BEIJING • SHANGHAI • HONG KONG • TAIPEI • CHENNAI

Published by

World Scientific Publishing Co. Pte. Ltd.

5 Toh Tuck Link, Singapore 596224

USA office: 27 Warren Street, Suite 401-402, Hackensack, NJ 07601

UK office: 57 Shelton Street, Covent Garden, London WC2H 9HE

British Library Cataloguing-in-Publication Data
A catalogue record for this book is available from the British Library.

AN ANTHOLOGY OF STRUCTURAL MORPHOLOGY

Desk Editor: Tjan Kwang Wei

ISBN-13 978-981-283-720-2
ISBN-10 981-283-720-5

Printed in Singapore by Mainland Press Pte Ltd

PREFACE

Which definition could be provided for the expression "Structural Morphology"? This interrogation was the first one when Ture Wester, Pieter Huybers, Jean François Gabriel and I decided to submit a proposal of working group to the executive council of the International Association for Shells and Spatial Structures. We were discussing about topics of high interest for us, topics related to shapes, mechanical behaviour, geometry of polyhedra and surfaces, design, bionics…We had no clear answer to our first question, but hopefully our proposal was accepted by the executive council chaired by Steve Medwadowsky. Ture Wester was the chairman of this working group until 2004, and I accepted to "take the baton" for some years. The year after, during the annual symposium of the association, I proposed to my colleagues to collect some of the papers that had been published for IASS events (symposia and seminars). Pieter Huybers suggested calling it "An anthology of structural morphology", and we agreed. This book gives our practical answer to our initial interrogation: while reading the thirteen chapters, a better understanding of "Structural Morphology" is possible. Several aspects of this discipline, which was introduced by famous pioneers like Leonardo da Vinci, Paxton, Graham Bell, E. Haeckel and more recently among others R. Le Ricolais, are illustrated by the contributions of many experts.

Next October our working group will hold its 6th international seminar in Acapulco. This seminar will be devoted to "Morphogenesis". A new era is beginning with younger members and contemporary problems. It is also my pleasure to know that members of the International Association for Shell and Spatial Structures and its President Professor John Abel are always interested in the works produced by our group. This book simply aims to be a small milestone on the road of structural morphology.

René Motro

Montpellier 27 June 2008

CONTENTS

CHAPTER 1

THE FIRST 13 YEARS OF STRUCTURAL MORPHOLOGY GROUP — A PERSONAL VIEW

Ture Wester

Associate Professor, Royal Danish Academy of Fine Arts, School of Architecture Inst. for Design & Communication, Philip de Langes Alle 10, DK-1435 Copenhagen K, Denmark
ture.wester@karch.dk

1. The Background

In this paper I will try to tell a short story about my personal experiences/adventures as chairman for the Structural Morphology Group, IASS working group No 15. I will describe the events as they are in my memory and my heart. This means that this paper is not a complete, even not a sketchy report of the many unique activities, papers, members, research etc. It is my own impression about the absolutely most important part of my professional life. I will go back in time and tell a little about why it is so: I graduated from the university at the age of 21 as structural engineer in 1963 and soon after became a teacher in Structural Design at the School of Architecture which had almost no tradition in research. My professor Jørgen Nielsen however was an exception and inspired my interest for research. After researching and teaching for 10 years in the topic of interactive dependency of shape and structural behaviour. These studies resulted in my discoveries of the geometrical characteristics and the necessary equation for the rigidity of pure structures in 3D and, most important, the concept about structural duality. At that time I had no contact with the international scientific world, but suddenly my "bubble burst" and thereafter everything developed extremely fast leading to the start of the group in 1991.

After a contact with Prof. Makowsky and Dr. Nooshin in Surrey, they encouraged me to write a paper for IASS'87 in Beijing. I went there and I was totally overwhelmed. I met wonderful people everywhere. Famous people I had read and heard about and, most important, people like me, researching on "useless" topics which made them isolated and maybe lonely in their university without sufficient local back up. As their research and the persons behind often were extremely fascinating and were dealing with the same interest of studying the intimacy between structure and shape, an obvious idea began to take shape: If these people were getting closer together they could fertilise and encourage each other and maybe even begin to collaborate on their "weird" research. The friendships made on these my first IASS events are still in the best of health.

2. The Beginning

For the rest of the paper I must apologise for any wrong or poor memory, any insult and all missing important information and mention - all unintended! In order to get a report over the activities please read the Newsletters of the group.

The IASS Working Group No 15 on Structural Morphology (SMG) was founded during the IASS Copenhagen Symposium in 1991. There was a relatively short period of planning before launching the group. It must be admitted that I never expected an international group with some of the world elite in the field as members. The first exchange of the idea was with Huybers on a bench in a park during an excursion in ISIS'89 in Budapest. One month later at IASS'89 Congress in Madrid, the "gang of 4" -Pieter Huybers, François Gabriel, René Motro, myself – established an action group with the aim to form a IASS working group. The specific name Structural Morphology was proposed by Michael Burt. Many of us had in the beginning a problem with the word "Morphology", as it appeared to be a biological term connected to flowers and dusty natural museums - and very far from engineering terms. But slowly it was clear that it was exactly the right word, as all other words as "form", "shape", "configuration" etc. indicated too narrow concepts: "Morphology" simply means "Study of Form" and is used in many sciences. While I was "captured" to be co-organizer for IASS'91 in Copenhagen, I had the real pleasure to work together with Mogensen, Støttrup-Andersen and not least Medwadowski, who became our President during that symposium. Medwadowski taught me a lot about IASS, international relations, and behaviour - and he supported the idea of a new working group on Structural Morphology. Based on an proposal from the 'gang of 4' backed up by a number of outstanding scientists, the group was officially formed during the Copenhagen symposium. The first SMG meeting took place September 5, 1991 in a small Danish pub 'Færge Cafeen', with Heinz & Maria Isler, Shikiko Saitoh, John Chilton, Pieter Huybers, René Motro, Michael Balz, Tony Robbin, Koji Miyazaki, Philippe Samyn, N.K. Srivastava, A.H. Noble, Wolf Pearlman, François Gabriel and me as founding members. Isler was, as always, very supportive and full of ideas. Just after the symposium, the gang met at my farmhouse at Møn to agree on a strategy for the future and afterwards was held a small (first?) seminar with paper presentations and discussions at the School of Architecture.

3. Group Activities

In the following chapters I have tried to convey some of my unforgettable experiences with my group. Not only professional matters but also social and artistic events during particularly our seminars will be discussed. I will try to describe these activities in some text and some photos. However, I have to apologise for missing description of important activities, experiences and events in the following chapters. This is not because of unwillingness, but simply poor memory or lack of awareness.

3.1.1. *SMG1 – Montpellier Seminar 1992*

This first Seminar had René Motro as the dynamic chief organiser. It was held at an impressive international level, very close to the standard of the IASS Symposia. Besides the traditional slide presentations, there were organized various workshops – or rather demonstrations – on e.g. Polyhedra and Architecture, Morphology and Computer Graphics, New Materials – New Morphologies, Tensegrity Systems, Surfaces and Membranes, Bionics, Foldables, Facetted Surfaces. Further four "Round Table" discussions on subjects as Natural Structures, Tensegrity Systems, Architecture Projects, Geometry and Architecture were arranged. These new types of activities were quite useful in an attempt to define the work and capabilities of the group, put a finger on the hot spots etc. A large impressive exhibition about the work of Le Ricolais was shown. The seminar is documented on video (unpublished) and in proceedings.

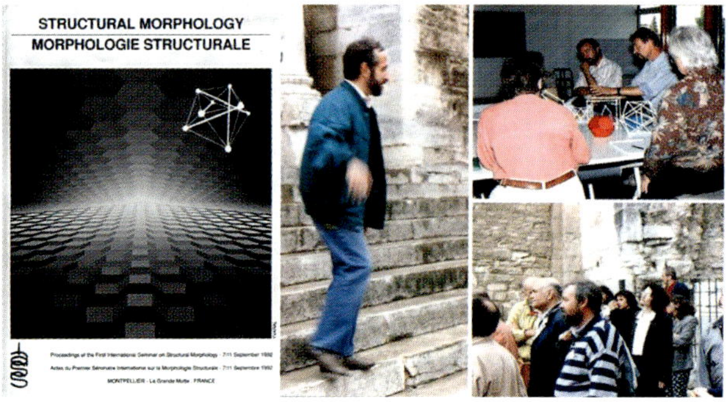

3.1.2. *SMG2 – Stuttgart Seminar 1994*

The two-day seminar was held in the best possible scenario for structural morphology: in the Balz/Isler concrete shells for the theatres at Stetten, and in the Institute for Lightweight Structures (IL), Frei Otto's world famous research institute at the University of Stuttgart in Vaihingen. Can anyone think of better venues for a seminar in Structural Morphology – it was really beyond imaginations. The organizer for the first day was the Michael Balz family. This day was very unusual, indeed. Taking advantage of the theatre stage, presentations were encouraged to be by physical models. The traditional type of slide-and-talk presentations was turned into performances with large models on stage – and even larger pneumatic models of airship garages outside. The most memorable performance was without any discussion Heinz Isler's demonstration of his model technique. It balanced constantly on the edge of success and disaster. There was smoke, heat, light, plaster, water, cracks, pumping air, shapes – and not least Heinz Isler in centre of it all, of course. It was clearly shown that model technique is a sensitive thing to transfer to a stage, but the performance turned out to be a tremendous success both from a professional and entertainment point of view. My student Cathrine Leth and I presented a 14 m2 foldable model. She had earlier been working with theatre performances and therefore turned the presentation into a carefully planed show with composed music and designed T-shirts for the occasion. I remember clearly that I should take off my jacket, shoes and socks because of my role as temporary support for the relatively heavy and unhandy structure.

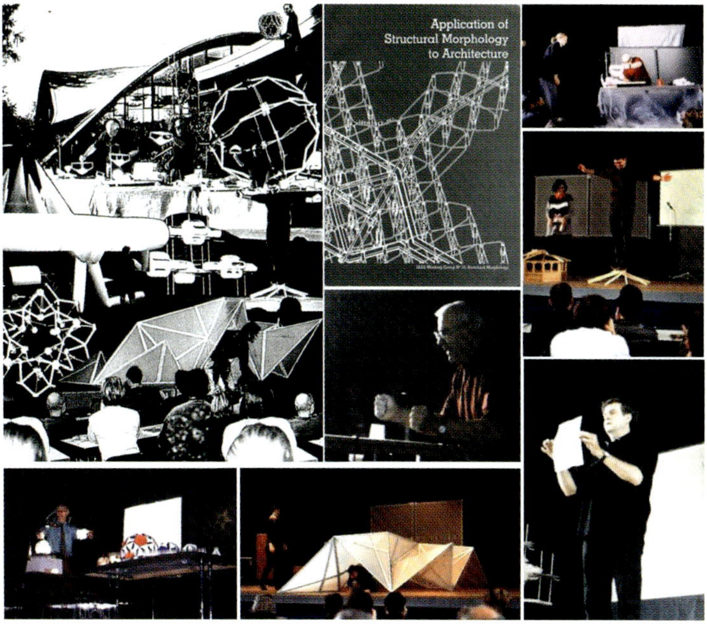

I was sitting besides Ekkehard Ramm on first row, and he whispered to me during my "striptease": 'I really hope that you know what you are doing!' After a long hectic day a dinner was served in the theatre, and a stage show was put together by a local amateur theatre group and some of the participants such as Chilton, Huybers, Kolodziejczyk, Servadio and Leth, and many more arranged improvised stage shows. Some of the events were luckily taped on video. This video has been edited and includes a number of very good takes from the day and the evening. The second day was organized by IL. This day was arranged in the traditional way of presentations, but in the special IL-atmosphere. I am probably prejudiced, but this seminar stands for me, because of its professional vitality and transcendent form, as the most interesting conference in my life.

3.1.3. *SMG 3 – Nottingham Colloquium 1997*

The conference was held at the University of Nottingham with John Chilton as the all-embracing main organiser. I remember him as almost collapsing from overworking just before the start of this memorable event (see the picture), but of course the perfect and relaxed host during the conference. In my mind the most innovative event was Jürgen Hennicke and students' "reconstruction" of Robin Hood's Sherwood Forest as six m high air-inflated tubes made from double layer sausage skin! Our President Stefan Medwadowsky, who participated, challenged the group by frequently asking about the definition of Structural Morphology? This question has been asked many times in all the 13 years and this is good because we should reflect about what we are doing. It is, however, in my opinion good that it has never been precisely answered. This would with no doubt bring some of our most interesting members in the periphery of our group – and what all this is about is getting these people in a synergetic contact with each other: Morphology means "Study of Form" and "Structural" is the same word with the same meaning as in the name of our association. The rest is up to oneself. This was the first time that our seminar was held in a university campus.

3.1.4. *SMG 4 – Delft Colloquium 2000*

In connection with Pieter Huybers' retirement from his university he organised a conference in Delft. As always with Pieter: everything must be perfect – and done in due time. This event was characterised by Pieter's calm and authoritative temperament. I specially remember Jürgen Hennicke and students air-inflated installations and Chilton's demonstrations of creating shells of clay. The technical tour went to the impressive Storm Surge Barrier – same size as two horizontal - and moving - Eiffel towers!

3.1.5. *SMG 5 – Montpellier Seminar (workshop) 2004*

This seminar closed my circle from 1991 to 2004 in Montpellier again with René Motro as the engine. He arranged not only the IASS Symposium but also a SMG workshop just before the symposium. It was very satisfactory to realize our improved skills to run workshops during those years This huge event went very well; even some of us (including myself) were very exhausted afterwards. I also began to feel that my enthusiasm by being the chairman had declined, and fresh blood should be infused. Also this time it was extremely satisfactory to see so many young people joining our arrangements. Our group seems to have a real attraction to the young generation of researchers. In many ways it was appropriate for me to resign from the SMG chairmanship leaving the responsibility to tireless Motro.

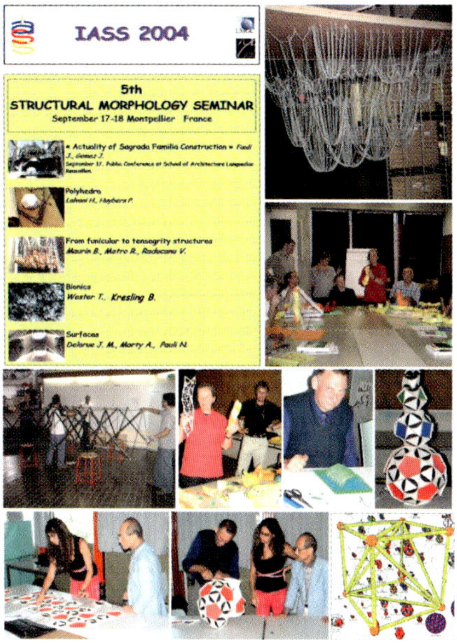

3.2. *SMG Newsletter*

Maybe the SMG Newsletters have been the most well-known and appreciated part of our group activities - worldwide as well as inside IASS. They have up to 60 pages with all kind of information: Detailed conference reports with paper abstracts, reports from our seminars, event calendar, reports from members, updated member list, book reviews, coming relevant events etc. have been included, and usually an original cover paper by a well-known researcher e.g. "A 'Monolytic' Granite Bridge in Beppu" by Mamoru Kawaguchi; "Metal Membrane Solar Disk Concentrators" by Jörg Schlaich; "Space and Structure" by Masao Saito; "Students Training in Form-finding and Analysis of Tension Structures" by Klaus Linkwitz; "Genesis of Structures, Structural Optimisation" by Ekkehard Ramm and Kai-Uwe Bletzinger; "Nexorades" by Olivier Baverel et al.; "A Sports Hall in the Shape of a Sword Guard" by Mamoru Kawaguchi; "Inflating Delft" by Jürgen Hennicke; "Adaptive Lightweight Structures" by Werner Sobek and Patrik Teuffel; "The Amsterdam Canopy: Designing and Prototyping a Computer Generated Structural Form" by Kristina Shea.

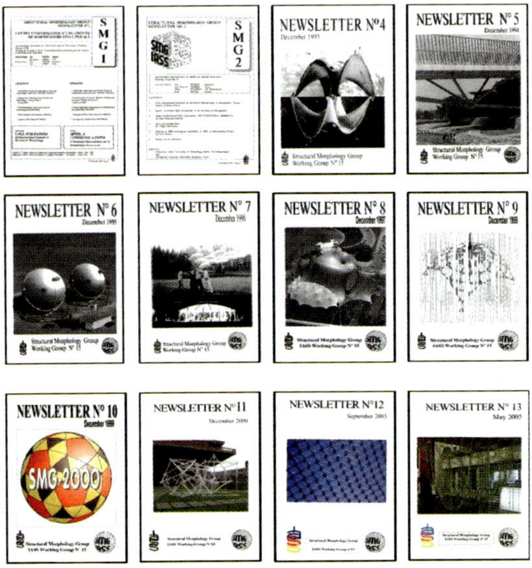

The extremely time consuming activity of editing of the Newsletter has circulated between the members of the "Gang of Four" while the financial support for the printing and distribution have been sponsored by the TU Delft and Ecole d'Architecture Languedoc-Roussillon, independently of IASS.

The Newsletter is the one and only way to get a real comprehensive report over the various activities of the group.

3.3. *SMG Activities at IASS Conferences e.g. Sessions, Meetings*

Since IASS'91 in Copenhagen we have at every IASS symposium had an open and very well attended SMG meeting, inviting participants to get information and join discussions on present and future activities. It was interesting to notice that the majority on these meetings were students and young researchers. In the recent years we also arranged sessions of our own at IASS symposia. At IASS'96 in Stuttgart there was a special afternoon dedicated to structural morphology, where Chuck Hoberman and Marek Kolodziejczyk were performing highly artistic demonstrations with structures.

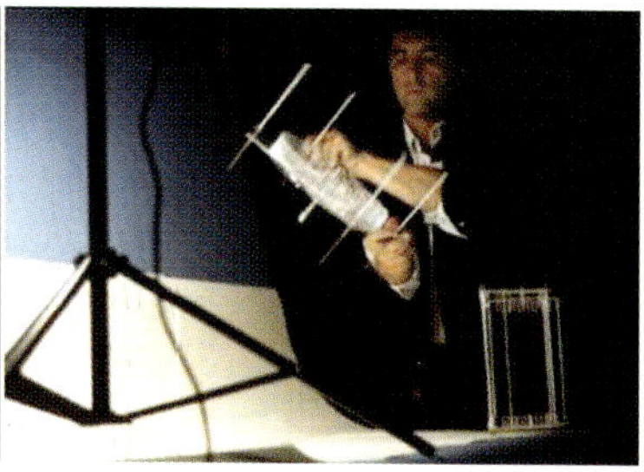

3.4. *The SMG Seminar Proceedings (see also chapter 3.1)*

1. Structural Morphology Proceedings of the First International Seminar on Structural Morphology. (ed. Motro & Wester). 479 pages. Montpellier, September 7-11, 1992.
2. Application of Structural Morphology to Architecture Proceedings of the Second International Seminar on Structural Morphology. (ed. Höller, Hennicke & Klenk). 255 pages. Stuttgart, October 8-9, 1994.
3. Structural Morphology – Towards the New Millennium Proceedings of the Third International Colloquium on Structural Morphology. (ed. Chilton, Choo, Lewis & Popovic). 375 pages. Nottingham, August 15-17, 1997.
4. Bridge between Civil Engineering and Architecture Proceedings of the Fourth International Colloquium on Structural Morphology. (ed. Joop Gerrits). 432 pages. Delft, August 17-19, 2000.
5. Fifth International Seminar on Structural Morphology held in Montpellier as a workshop. There exists no proceedings from this event but a short report is available in SMG Newsletter No 13. Montpellier, September 17-18, 2004.

4. Further Recent SMG-Relevant Reading

Publications with major relevance to Structural Morphology and written by a SMG member. I apologise my poor memory and lack of knowledge for missing items.

1. The International Journal of Space Structures edited by H. Nooshin and Z. Makowski. Most of our members have written articles and

other contributions to this unquestionable the best independent international journal with structural morphology as one of the major fields of interest. Also it is not a coincidence that half of the editorial board and the two chief editors are SMG members.

2. The International Journal of Space Structures, special issue on Geodesic Forms. Vol. 5, Nos. 3&4. 378 pages. Guest editor Tibor Tarnai. Contributions by SMG-members: Wester, Rebielak, Lalvani, Huybers, Miyazaki, Miura, Motro.
3. The International Journal of Space Structures, special issue on Tensegrity Systems, Vol. 7, No. 2. 90 pages. Guest editor René Motro. Contributions by SMG-members: Motro, Loeb, Hangai, K. Kawaguchi, Pellegrino.
4. The International Journal of Space Structures, special issue on Deployable Structures, Vol. 8, Nos. 1&2. 147 pages. Guest editor S. Pellegrino. Contributions by SMG-members: Pellegrino(2), Miura, Escrig, Verheyen.
5. The International Journal of Space Structures, special issue on Morphology and Architecture, Vol. 11, Nos. 1&2. 274 pages. Guest editor Haresh Lalvani. Contributions by SMG-members: Tarnai, Lalvani(2), Gabriel, Pearce, Burt, Miyazaki, Robbin, Loeb, Huybers, Verheyen, Wester, Motro, Rebielak.
6. HyperSpace. Japan Society for Hyperspace Science, edited by Koji Miyazaki. Frequent contributions by SMG-members: Burt, Lalvani, Miyazaki, Robbin, Tarnai, Wester.
7. Engineering a New Architecture (1996) by Tony Robbin, Published by Yale University Press 1996, 138 pages.
8. The Periodic Table of the Polyhedral Universe by Michael Burt. Published by Technion, Israel Institute of Technology 1996, 145 pages.
9. Beyond the Cube edited by J. Francois Gabriel. Published by John Wiley & Sons, Inc. 1997. 510 pages. Contributions by SMG-members: Loeb(2), Saitoh, Huybers, Motro, Wester, Nooshin, Hanaor, Lalvani, Robbin.
10. Geometric Forms in Architecture (in Japanese) by Koji Miyazaki. 2000, 191 pages. Reference to many SMG-members.

11. Heinz Isler by John Chilton. Published by Thomas Telford Ltd. 2000, 170 pages.
12. Space Grid Structures by John Chilton. Published by Architectural Press UK. 2000, 180 pages.
13. Story of Space and Structure – Structural Design's Future (in Japanese) by Masao Saitoh. 2003, 271 pages.
14. Tensegrity – Structural Systems for the Future by René Motro. Published by Hermes Science Publishing, UK. 2003, 236 pages.

5. Tsuboi Awards

Around 10 of our members have received the prestigious Tsuboi Award, and around one third of Executive Council is a SMG member, even one vice-president. This indicates the high standard and dedication of our members.

6. Research and End Notes

It is an impossible task to report on this topic as our members have through the 13 years written many hundreds – maybe more -of excellent and often innovative papers presenting completely new ideas, both for SMG-seminars, IASS Symposia and for other conferences, international journals, books, exhibitions, TV, and all other possibilities for publishing. I admire the energy, skills and hard-working members for bringing the concept of this important and fast growing field out to every corner the scientific world of engineering and architecture – and to have made the Structural Morphology Group - the IASS Working Group No 15 well-known and respected all over the world.

Acknowledgements

I owe the Gang and the many loyal members everything for the success of the group. I have many times claimed that the success of this group is not defined by its leaders but because of the synergetic effect created when so many outstanding professionals – researchers, practitioners, artists, architects, engineers etc. – work together, inspire, support and constructively discuss with each other. Thank you for the friendships which met me across all borders that evolved inside the group. The friendly atmosphere, the high spirit and professional level mixed with artistic touch has been unique, everything because of the members contribution as individuals or in groups. I bet that all who have participated in our seminars will agree that the mixture of warm and friendly spirit, and high professional standard, has been unsurpassed. It is my sincere hope that this precious quality will remain in the future. It has been a shear pleasure for me to have been the chair for this particular group from its birth to teenager. I think many of the past activities will continue in the future but times changes - sometimes very fast – especially for a "teenager". SMG has recently included a very energetic subgroup FFD (Free Form Design) mainly based on young researches from TU Delft, who could fertilise our group with new energy, new topics, young people – besides, the FFD topics seems to be right in the centre of Structural Morphology – whatever the definition is! As many of you already know, I left the chairmanship of health reasons, lack of sufficient support from my university, but also my feeling that young blood should be infused to the leadership. I am convinced that René Motro is the best qualified to handle this process. The photos I have used in this paper are taken by group members, but I don't remember who has taken which photo, so I thank all who through the years have send me these memorable photos.

CHAPTER 2

AN APPROACH TO STRUCTURAL MORPHOLOGY

René Motro

Laboratoire de Mécanique et Génie Civil. Université de Montpellier 2. Groupe de Recherche et Réalisation de Structures Légères pour l'Architecture

This communication aims at following several lines of questions concerning structural morphology. This is re-positioned in the general context of the design of construction systems to restore the full organisational sense to the word 'structure'. The system proposed makes it possible to classify design parameters in four categories: forms/ forces/ material and structure. All the data related to the four parameters are subjected to constraints. The system which is the subject of the design must among other things meet mechanical criteria. The problem of design can thus be handled by ordinary systems optimisation modelling. The position of structural morphology in this system is at the interface between the parameters 'form' and 'structure'; it expresses their interactions and meets the requirements of the material and the balance required. A number of landmarks give a glimpse of the potential that can be hoped for from research in structural morphology. They are chosen from fields in which they have already given results but without exhausting the potential: that of stone cutting and that of geometrical studies in which stress must be laid on both topological and dimensional aspects. More generally, structural morphology can draw on bionics on condition that this is not limited to surface of phenomena and that questions are raised about the underlying principles behind them. Some classes of systems cannot do without interactions between form and structure. This is the case of "equilibrium forms" illustrated by tensegrity systems and membranes. It can be considered that the form expresses the meaning of the structure, that is to say the meaning of the choice of organisation of matter decided by the designer. In fact it goes beyond the meaning of the structure to show that of architecture, and the final coherence of meaning should be sought.

Introduction

The title of this communication fixes the limits - an approach rather than complete coverage. It cannot be otherwise in a seminar where one of the aims to draw together scattered items to give impetus to a field which has existed since man started building structures but which is not established as a discipline. We propose a method of approach which cannot be considered as modelling but which makes it possible to examine in the same way several landmarks in the past and present state of the building sciences before asking questions on the meaning which can be given to research on structural morphology with a view to enriching architectural design.

1. Structural Morphology

1.1. *Structure and System*

1.1.1. *Structure*

The word "structure" expresses a concept which has not been the subject of many controversies in spite of the inaccuracy associated with it. The very aim of our work required a more accurate approach to the concept. Systems theory and its definitions shed light on the concept of structure.

'Struere', meaning 'build' is the Latin root of the word 'structure'. Vitruvius used this meaning of the word in his treatise on architecture (27-23 B.C.). For Vitruvius/ 'structura' was brick or stone and mortar masonry. This means that the word had a building connotation from the outset. 'Archeology' of the idea of structure would have to establish the date and context of the first use of the word to indicate not just a mass of inert masonry but the building itself with its own order - that of a construction with both mechanical and functional determinants.

Architectural metaphor has a somewhat unsuspected role in the archaeology of structural thinking in general. It has been the source of numerous models, generally of the mechanistic type based on the distinction between form and structure inherited from Viollet le Duc.[17] His main contribution was the study of the science of construction, of structure, which governs all the formal and decorative features of Gothic architecture. Viollet le Duc thus showed the need for structural analysis of architecture, implying the abandoning of the strictly descriptive point of view. This model has been used in recent years in fields such as linguistics which, in the structuralist field is based on the architectural metaphor.

"This structural analysis is first of all a desire to consider the architectural phenomenon in terms of "Systems", which are linked and coherent to varying degrees, and in such a way that a change made to any part can but be felt in other parts of the constructed organism.

This is an affirmation that architectural syntax, like the mechanisms of articulated language, is not reduced to a combination, a coordination of forms of equal value, but that it covers a ranked organisation of constituent units, the latter being disposed according to a strict order but whose subordination is variable.

It is finally my intuition that, if the architectural phenomenon has a meaning, this should not be sought at component level but in the System itself".[4]

1.1.2. *System*

Systems and structuralist movement theoreticians[1] have drawn much of their vocabulary from the language of architecture. The problem for defenders of structural thinking is to go beyond the mechanistic model. However, builders -engineers in particular - should take advantage of progress in the deepening of method in the archaeology of which its objects have a decisive position. This point led to this description of the results of systems epistemology performed with a multi-field approach by researchers who have deepened concepts such as structure, system and form. Better understanding of terms, most of which are common in builders' vocabulary, can open up new pathways in "structure design" related to morphology, the subject of this paper.

The notion of "system", as defined by P. Delattre is a first stage:

"The notion of system can be defined very generally by saying that a system is a set of elements which interact between each other and possibly with the external environment."[6]

One of Delattre's axioms concerned elements: the behaviour of an element is related to the role that it plays in the System, i.e. its properties. Comparison of several elements involves comparison of their properties, enabling the definition of reference categories. Elements are material or abstract and defined by two types of characteristics:

- qualitative characteristics,
- quantitative characteristics.

The qualitative predicate necessarily precedes the subordinate quantitative predicate. The latter represents the relationship to a certain frame of reference.

Once the characteristics of the elements have been fixed, it is possible to define the "reference categories" grouping all the elements with the same characteristics. In the case of built systems, the number of elements in each category is reduced to one when the spatial position is one of the characteristics of the definition, since a geometrical point can only be occupied by one material point.

The effect of interactions between elements is shown by the movement of elements from one reference category to another (which takes place continuously for example in beam deformation). The elements are either modifiable (modification of the quantitative data of all or some of the elements) or transformable (disappearance or appearance of qualitative characteristics).

Following the interactions to which they contribute, the elements of a system may undergo transformations and run through various classes of equivalence successively. We refer to a transformation system in this case. Static systems can be considered as a special case of transformation Systems without change between categories. The relations between categories express the stresses between the elements of the system and which are responsible for its static equilibrium.

Two types of relation can be envisaged in this case:

* order relations established using comparison of the quantitative characteristics for a given qualitative characteristic,
* topological relations of continuity and proximity.

Delattre used this as the basis for listing the parts of a full definition:

* affirmation of the existence of categories and relations between categories,
* the kinds of relations between categories (order, topology),
* the kinds of elements in the different categories: listing **of** their characteristics,
* the number of elements in each category,
* the analytical form of the expressions linked to the relations between categories (assembly of the characteristics and the corresponding numerical values).

1.1.3. *Structure and System*

"Structure" is used in the broad sense here: "the manner in which the parts of a whole are arranged". This definition serves only to affirm the existence of an Entity, parts and organisation without establishing the nature of the elements. This concept is therefore included in that of system whose dynamic aspect is inherent in the very nature of the elements which can alone account for the interactions.

It is then possible to apply a reduction process to the elements of definition of the system to extract a definition of its structure. If one keeps to the first two elements of definition given by Delattre the nature of the elements is not specified. He called this a Relational Structure.

This is the meaning used in mathematics for group structure of an ensemble, for example.

A second level of definition is obtained by adding the third element of definition of systems - the list of the characteristics of the reference categories grouping the elements. The definition reached corresponds to the Total Structure of the system. Elements whose characteristics are specified may themselves be arranged in systems; the total structure implicitly contains the structure of subjacent levels at the description level chosen, which does not contain the relational structure.

It is important to observe that onc cannot talk in terms of relational or total structure without defining the system to which they are related. The level of description chosen must also be specified. The subject being "design in structures", research attention was paid to both the total and relational levels.

1.1.4. *Conclusion*

Going further into the notion of structure outlined in this part would be meaningless if the subsequent formalisation were aimed at replacing the models well known to engineers and technicians. The aim of this approach is to re-establish reflection on the design of structures by removing ail unrelated information. It then remains to be seen how this can be useful for structural design in the context of its close relation with architecture.

1.2. *Form and Structure*

An attempt at defining the concept of "structure" soon leads to simultaneous consideration of that of "form". The two concepts are closely related and their history displays inverse evolution. Definition of one often makes reference to the other. Thus, the French Larousse dictionary defines "structure" as follows in the psychology section:

"...an organic set of forms which, according to some psychologists, is perceived directly before each detail is isolated".

On form: "Form theory, a theory considering the perception of a set of organised structures before the details are perceived and which affirms in all domains the influence of the whole on its constituent parts."

Structure is defined as the "manner in which the components of a whole are assembled". The definitions proposed for the concepts clearly show the existence of a whole and its parts; these notions can **be** related to that of System.

In fact, the evolution of the two concepts can be clarified by noting that both are used in a limited sense to show the spatial existence of the object in question. Today, the concept of "form" is mainly used for the limited sense of spatial configuration and that of "structure" is used in a broader sense. It is known that the Ancients did the opposite - or at least "form" was the broad concept for them. The word "structure" replaced "form" little by little, leaving it only the limited sense.

The confusion still exists for example in the translation into French of "Gestalttheorie" / "Théorie de la Forme"[7]; an accurate translation would have used the word structure, although the theory effectively concerns our perception of an object. In line with modem comprehension of the two terms, we have chosen to consider the term "form" in its limited sense of spatial representation.

1.3. *The Design Process*

1.3.1. *Structures and Constructions*

Although, as is stressed above, the vocabulary of Systems theory owes much to architectural metaphor, study of construction systems and constructions has certainly not drawn all the advantages that can be expected from a systemic approach. The term "constructive system" is used in the everyday sense here without giving it a specific character making it the field of the architect or the engineer. The difficulties in establishing or abolishing boundaries between constructions which may be attributed to one camp or the other are well known.[2,13] Nevertheless, both use the word "structure" instead of "system" to signify what we shall call the "resistant" part of their constructions for want of a better word. If one keeps to the definitions in the preceding paragraphs, one takes the part as being a whole without separating the nature of the principle of organisation from its system.

Complements and adjectives are added to the term "structure" to define construction Systems in specific cases. This is the case of the phrase "reticulate structures", which provides information on the nature of the relations between elements; this information can be expressed by analytical formulae, which comes down to describing the system. The expression "metal structures" and all those concerning materials only specify one of the qualitative predicates of the elements. The same applies to names like "stressed structures" or "compressed structures" which describe a system characterised by a stress state.

Confusion between terms is not a problem in a cognitive study or a design for which there is complete knowledge of the system. The situation is different in design problems which require the creation of a system. It must be possible to clearly define and understand the share of each of the elements defining the System in order to achieve an "optimal" construction.

Although research on the characteristics of the elements (especially the determination of a large number of those concerning the constituent material in "rheological" studies) and work on elaborating cognitive and forecasting mathematical models have been discussed at length, it has been based on principles of organisation which are rarely questioned. "Relational" structure as defined by Delattre is accepted without evaluation of its effect. Topological relations and relations of order are nevertheless of great importance in the properties shown by certain systems. Interesting results can be shown in particular on the basis of the relational structure of a system only; these results show intrinsic properties, that is to say features which are independent of the environment of the system.

1.3.2. *Design of Structures*

What methods should be used? The question does not require a single answer. It is first necessary to find the best way of identifying the problems and to show the dominant requirements. Although it is conceivable to draw up correspondence between a matrix of the categories of problems and another of categories of solutions/ a different approach would be to classify the solutions horizontally or vertically. In the first case, existing solutions are applied to the case covered or by pushing back the known limits. In the second case/ there will be a change in level through the design of an innovatory solution; this is when the weight of the influence of the dominant models must be accurately evaluated. The completion of innovatory solutions requires much more energy and conviction and structural morphology studies form a good basis.

1.3.3. *Modelling the Design Process*

As the construction system is the purpose of design, the formal framework of aid for design must be defined. It consists of the following items:

- definition of the properties expected
- list of constraints
- list of the design parameters.

The study framework thus defined is the description of a System optimisation problem.[15] The characteristic of the approach lies mainly in the listing and processing of the design parameters/ which consist of four families:

- structure (in the relational and/or total sense described above)
- form, considered as the projection of the entire structure in geometrical space,
- force, consisting of all the mechanical notions related to the design of constructions (actions, stresses and strains, etc.),
- material, study of mechanical, physicochemical properties, study of be behaviour.

There remains, of course, the task of listing the properties expected and the constraints. The relation with architecture is effected at this level. It cannot be considered that the only meeting point is that of form, even if carries considerable weight. It would seem that the four families of parameters mentioned in connection with the "design of structures" is a sub-set of those involved in architecture. Each contributes to the general "meaning" of the result.

For example, if the parameter "structure" is characterised by the notion of rhythm, this must not conflict with the rhythm induced by the circulation planned. Work in this field is in the very early stages in our research, but the aim is to combine different levels of structures (mechanical/ architectural, symbolic, etc.) with the relational meaning of the term to create synergy and not contradictions which would weaken the design itself. The modelling of the problem in this way does not provide answers to ail queries but sheds as much light as possible on the points which can be used for the development of research in structural design.

1.4. *Structural Morphology*

The position of structural morphology in the proposed modelling in the preceding paragraph is illustrated by the diagram in the figure opposite.

It is in fact the direct relation between the study of form and structure extended to cover the relational sense. This relation is affected by the behaviour of the material and by the need to ensure the static (and sometimes dynamic) equilibrium of the system S being designed. It is not complete theoretical modelling but a method of approaching the problem. Other parameters must be considered, and especially those related to the technological facilities available for the construction System. The cost of construction and operation is an important factor in evaluation.

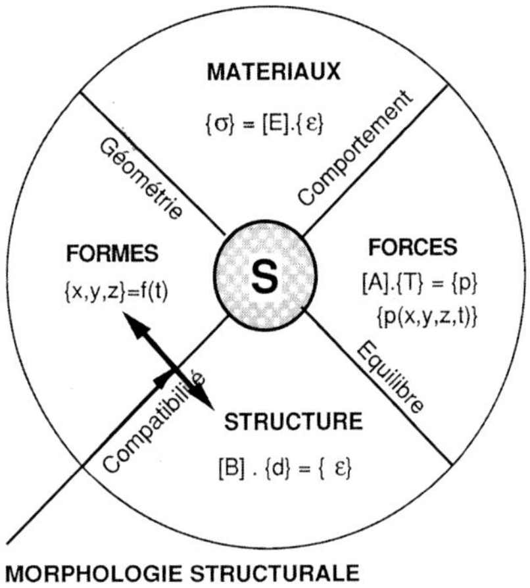

Fig. 1. Conceptual scheme.

2. Landmarks

The effect of the form-structure relation as described above is present in almost all construction systems. The importance of the role of the two other parameters depends on the case. Four examples are chosen as "references" and described briefly in the light of the conceptual procedure.

2.1. *Stone Cutting*

Stone cutting plays a very important role in the history of building. The material is characterised by a dominant property of compressive strength and gravity is the determinant element in all the actions applied. Careful stereotomy is required to ensure static equilibrium among other things. The form of the elements governs their dead load and the structure conditions all the stresses. Setting aside purely vertical Systems in which single-direction compressive strength affects the material almost uniformly, we mention two solutions in which structural morphology plays a major rôle. Observing the limits imposed by the bending of stone material, builders devised other solutions whose common feature is the use of elements which are small in relation to span in a structural assembly which provides equilibrium. Whereas the transition from straight beam to arches with arch stones is well known, the solution used in the case of the lintel for the tympanum of a Gothic gate at Alba lulia (Romania) is less so (Fig. 2). Large blocks are not required and the form given to the elements is stable and original.

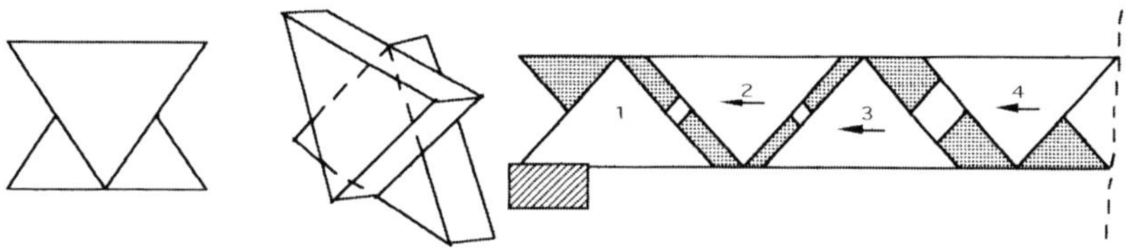

Fig. 2. Stone cutting principle (Alba Iulia).

2.2. *Geometry*

Geometry plays a central role in the design of form in the broad sense. It is related to the partitioning of space, whether this is discrete or continuous. Designers make extensive use of polyhedral geometry in the first case and the geometry of surfaces in the second. The two approaches meet when the problem is not one of partitioning space but of handling special surfaces, such as geodesic domes. Examples can be found both in the work of Graham Bell at the beginning of the century and in the steel framework of the Zeiss planetarium in lena in Germany in 1923.

Researchers like Le Ricolais and Fuller have taken the question further. The former established the basis for space-frames which were developed very widely; his observation of Radiolaria went beyond the purely geometrical stage to cover examination of the relation between networks and induced forces. In his studies on *"Structures comparées en deux et trois dimensions"*,[11] he drew a distinction between several classes of square, triangular and light or dense hexagonal mesh networks. Basing his work on Mayor's research on graphic static, he then revealed the duality of geometry and forces by stressing their dual topology (*Méthode Image*,[12]); he stressed here the need to examine the constitution of the initial pattern and its multiplication. In contrast, Fuller's contribution in the field of geodesic domes was directed towards what one could call geometrical regularity without explicit reference to the mechanical implications of the choices made. Until recent years it was almost obligatory to have the smallest possible number of different components in order to meet industrial and construction constraints; adaptation to stresses was and still is handled by modifying the straight sections of the elements.

Today, these imperatives are less important thanks to progress in computer-aided manufacture. Several questions are now of interest: should one keep to simple geometrical regularity in the dimensional and angular sense or should topological aspects be stressed? Regular and semi-regular polyhedral geometry is certainly an interesting basis, but it does not cover the mechanical aspect of the behaviour of structures in a given environment of external actions. Under these conditions, the problem of the limits of useful investigation in polyhedral geometry is raised. Facilities such as Formex algebra on the one hand and homogenisation on the other will certainly make it possible to model the real behaviour of the systems on the basis of the constituent pattern and the relational structure rule chosen. In fact, it is known how to reconstitute a complex geodesic dome using a very small number of data (Schwarz's triangle). Discussion of the question must be very broad.

2.3. *Bionics*

Structural morphology draws heavily on bionics, which can be considered as "the science of systems whose functioning is copied from that of natural systems or which display specific features of natural Systems or which are analogous". This was how Y. Coineau's reported the definition of bionics given by Major Jack. E. Steele in the first congress on the subject in Dayton, Ohio in 1960.[3] In construction systems, the work by Le Ricolais mentioned above has enabled the development of space-frames, and extremely important work was performed by Frei Otto and the researchers at his institute in Stuttgart. However, attention must always be paid to the intellectual procedure chosen in an approach to bionics. A simple "formal" copy in the strict sense of the word generally has no mechanical or other substrate. As Laënneck noted, "You can sit on china eggs for as long as you like but you'll never get any chicks". When Calatrava designed a structure deduced from the shape of birds' wings he did not claim to give it the power of flight. The statement was more symbolic. Leonardo da Vinci's approach to the functioning of bats was very different and was aimed at designing flying machines. Querying nature usually generates new solutions not through simple copies but through understanding the fundamental principles of operation. It may thus be erroneous to study a natural phenomenon without covering the whole of this phenomenon. Modelling the structure of trees is not the simple application of structural mechanics to a material, whether considered as anisotropic or heterogeneous; the mode of growth is important. The evolution in time of all living material must be taken into account so as not to make meaningless analogies.

The environment in the broad sense affects design in the same way. The known analogy between the bone trabeculae in the head of a femur with the isostatic lines associated with a classic loading must be examined in other environments. Study of the evolution of bone structure in spacemen reveals what is reported to be a significant modification or reduction in the trabeculae.

The principles underlying form must be discussed. D'Arcy Wentworth Thompson's work on honeycombs revealed geometry which can be related to the partitioning of space by rhombic dodecahedra. On a different scale, some recent results on agar gels reveal the same type of geometrical layout. Agar was found for the first time by Payen in a substance called "Chinese moss" and it is found in some seaweeds (Java weed and common *Gelidium*). Gelatine and agar can be heated to form aqueous solutions which cool to form more or less elastic gel. It is interesting to note that the resulting geometry divides space into a polyhedral layout with the matter forming the ribs. This is caused by Van der Waals forces resulted intermolecular attraction. Matter distribution is therefore not identical to that of honeycomb, where it forms the polyhedral faces, but the two phenomena nevertheless possess relational structures of the same type and their origin should be understood. This example alone should encourage prudence on the part of all those who venture into bionics: the meaning behind the appearance should be sought. Bionics is at a crossroads with several other sciences and requires both an open mind and a rigorous approach.

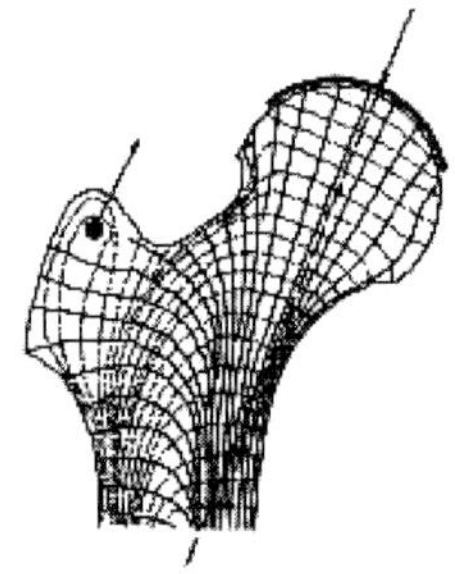

Fig. 3. Cross-section of a bone.

2.4. *Equilibrium Forms*

The diagram (Fig. 1) in §1.4 places structural morphology as a link between form and structure but this does not mean that the two other parameters are without effect. There are construction systems in which form depends directly on the forces applied and their equilibrium. The stiffening of these systems in a definite form is the result of the forces applied. There are two categories of equilibrium form systems:

- "funicular" systems (in the broad sense of the term) in which form depends on the external system of actions applied. The denomination results from the equilibrium of cables subjected to individual actions. The best-known application is that of Gaudi in "Colonia Guell". He took advantage of the duality of compression and tension and defined compressed systems by inverting entirely tension systems, in this category, forms and forces are biunivoque; each system of distinct forces is associated with a different form.

– "self-stressing" systems. Stiffening results from a system of internal stresses in static equilibrium. The equilibrium form is not effected by external actions - if only in relation to strain resulting from the nature of the materials. The bar and cable systems popularised as "tensegrity" Systems are in this category, as are "textile membranes" and "cable networks" whose equilibrium requires complementary structural elements. The design of these equilibrium forms requires research on form. In the two cases illustrated in Figures 4 and 5 this is based on a grid constructed using Formex algebra. A numerical research method (dynamic relaxation for tensegrity Systems, force densities method for membranes) gives the equilibrium form Ref. 15.

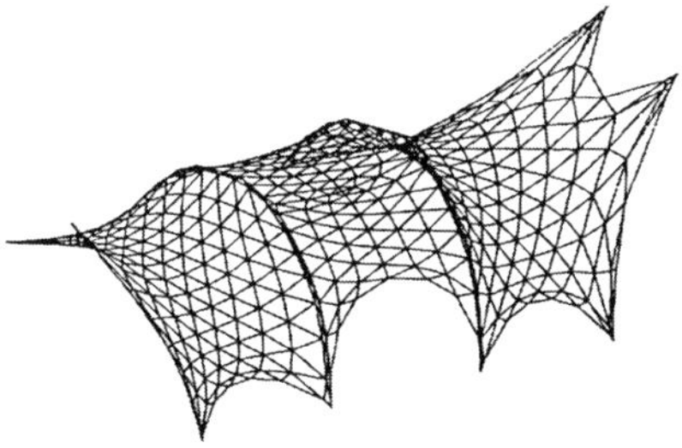

Fig. 4. Membrane form finding.

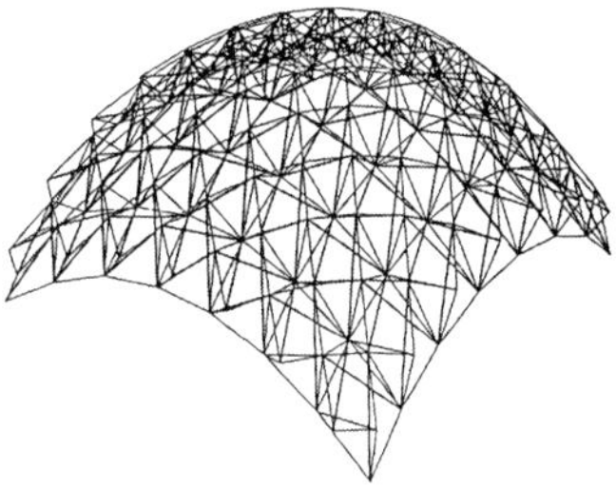

Fig. 5. Double Layer Double Curvature Tensegrity system.

3. Form and Meaning

3.1. *Structural Morphology and Art*

The links between structural morphology and art can provide information for designers. A few examples show the richness of this type of approach, which is well illustrated in *Gödel, Escher and Bach.*[8]

The work of the sculptor Kenneth Snelson on tensegrity systems enabled Fuller to publicise them widely, but the relationship between human investigations is also shown. The structure of certain paintings strengthens the expression. This is the case of Cezanne's "Les

Baigneuses" strongly marked within a triangle. Ucello made full use of structural morphology in his three paintings of the Battle of San Romano. The composition enabled him to add several dimensions to those usually shown in a painting on a flat surface. The play of the lances gave him access to time and shows the dynamics of the situation through subtle orientation of the lances of the attacking and attacked horsemen. The form shows the underlying meaning without clouding it.

Painters showing particular interest in the notion of structure include Paul Klee, whose investigations went furthest. His study of trees and their growth is exemplary. It is in a work with the significant subtitle "Theoretical order of the means of creation related to study of nature and the constructive pathways of composition".[9]

3.2. *Order and Harmony*

"All the orders in the world, not only the Greek which is particularly conscious, but also of the ancient East and of the American Indian, are spiritual" wrote Kerenyi. They invite the drawing up of a harmonious plan. The concept of harmony is essential for establishing order. It expresses all the relations between the revealed and the non-revealed, between the whole and the parts of the whole. The clearest illustration is that of the division of a segment considered as a unit into mean and extreme ratio. Given a segment AB, seek a point of division C so that the new segments are in harmony with the initial segment. The ratio which is the solution to the problem is the golden ratio. Unfortunately, the name gives it a "magic" aspect whereas it is one of the basic principles of symmetry in the Greek sense of the term. The result is harmony in the sense of coherence between the whole and its parts. The mathematical translation is only a result of the intention and shows its meaning. This division is such that the small segment is to the medium what the medium is to the large one. It is not surprising to see this construction when an icosahedron is generated from a tetrahedron.[14] It is in fact another form of duality. Two figures or two concepts are dual; what the first is to the second the second is to the first in such a way that a twice performed dual transformation returns to the origin. Two dual elements are in balance and it is more appropriate to talk in terms of equilibrium action for example on compressive and tensile stresses in reticulate Systems.

3.3. *Form and Meaning*

The "form" of construction systems is closely related to the relational structure and shows its organising aspect; it bears the meaning that the designer intends to give his project. It is useful to pay attention not only to the pertinence of the project but also to the permanence of its expression. However, the project is not usually the engineer's or the surveyor's but that of the client. The constraints to be met and the functions to be performed are not only mechanical or plastic. Its meaning must be closely related to that of architecture and its imperatives. It can thus be understood that the structure described here is that of a subsystem which, in harmony with others, should lead to a meaningful construction. The part cannot exist without the whole. Study of structural morphology cannot ignore its context; it must be in its position.

Conclusion

This discussion of structural morphology is not the definition of a theoretical position but a query addressed to everyone to open up pathways and avoid dead ends. We have set out an open intellectual "diagram" and mentioned a number of references. Structural morphology is an interface science, a difficult field in which superficial analogies can rapidly lead to serious mistakes. Rigour in investigation and discussion with specialists in the fields concerned are both necessary. This can lead to giving a meaning to new design solutions and to full expression in constructions.

References

1. Auzias J.M., "Clefs pour le structuralisme". Seghers. 1971.
2. Billington D.P., "The Tower and the Bridge". Basic Books. New York. 1983.

3. Coisneau Y., Kresling B., "Les inventions de la nature et la biunique". Hachette. 1987.
4. Damish H., "Les fonctions de l'Ordre Classique", in P. Francastel "L'Urbanisme de Paris et de l'Europe". Colloque de l'E.P.H.E. Paris. 1969.
5. Delattre P., "Système, Structure, Evolution, Fonction, Evolution". Essai d'analyse épistémologique. Maloine Doin, éditeurs. Paris. 1971.
6. Delattre P., "Séminaire Interdisciplinaire du Collège de France : Structure et Dynamique des Systèmes". Maloine et Doin. Paris. Page 23. 1976.
7. Guillaume P., "La psychologie de la Forme". Flammarion. Paris. 1979. Page 5.
8. Hofstadter D., "Gödel, Escher, Bach, les Brins d'une Guirlande Eternelle". InterEditions. 1985.
9. Klee P., "Histoire Naturelle Infinie". Dessain et Tolra Editeurs. 1977.
10. Lichnerowicz A., Gadoffre G., "Genèse et objectifs des séminaires interdisciplinaires". Avant-propos au séminaire interdisciplinaire du Collège de France : Structure et Dynamique des Systèmes. Maloine et Doin. Paris. Page 7. 1976.
11. Le Ricolais R., "Structures comparées à deux et trois dimensions". 1946. Repris dans Techniques n°9 et 10.
12. Le Ricolais R., "A la recherche d'une mécanique des formes". Conférence donnée au Palais de la Découverte. 1965.
13. Motro R., Rapport sur le Symposium de l'IASS "Development of Form". Morgantown. 1978.
14. Motro R., "Formes et Forces dans les Systèmes Constructifs. Application au cas des Systèmes Réticulés Spatiaux Autocontraints". Thèse d'Etat. Montpellier. 1983.
15. Motro R., "Conception en Structures". Séminaire sur les Processus de Conception. Paris. 2-3 avril 1992. Pages 212-222.
16. Vitruve M.V.P., "De Architectura libri decem" 27-23 a.j.c.
17. Viollet Le Duc, "Dictionnaire raisonné de l'Architecture Française ». 10 vol. Paris. 1854-1868.

CHAPTER 3

THE STRUCTURAL MORPHOLOGY OF CURVED DIAPHRAGMS — OR THE STRUCTURAL BEHAVIOUR OF FLORAL POLYHEDRA

Ture Wester

Associate Professor, Royal Danish Academy of Fine Arts, School of Architecture, Copenhagen

The publication by Michael Burt *The Periodic Table of the Polyhedra Universe* [Ref. 2] interpret Euler's theorem on polyhedra in three dimensional space to its extreme, and shows by this that the classical plane faceted convex polyhedra just are a tiny part of the full spectrum. Burt shows that polyhedra of any genus - including infinite polyhedra and so-called *floral polyhedra* where some are not imaginable with plane facets and straight edges - are included in the Euler-polyhedral universe. *Structural Order in Space* [Ref. 5] is investigating the structural behaviour of any plane faceted so-called conventional polyhedron, and deduces the necessary stability equation for structural configurations stabilised by shear forces alone – i.e., pure plate action - and it reveals the profound interrelation between lattice and plate structures - the structural duality - which follows the well known geometrical duality. These statical relations for shear-stabilised polyhedra and the structural duality turns out to be valid for any plane facetted configuration and they can be extended from the level of *topology/stability to metric geometry/magnitude of forces and elasticity*, as shown in [Refs. 6 & 7]. It is therefore a necessity to try to increase the statical considerations to cope with the full understanding of Euler-polyhedra as described by Burt where the floral type is the normal and the plane faceted type is a rare specialty. The present paper is an attempt to approach the full spectrum of polyhedra as suggested by Burt.

1. Basic Stability in 3-Dimensional Space

The following statements on stability are the necessary but not sufficient requirements.

✧ For lattice structures:

A node is stabilized by three supports.

Two nodes linked by one bar are stabilized by five supports.

Three nodes linked by three bars - and any other 3-dimensional internally stable lattice configurations - are stabilized by six supports.

As stated by Moebius-1837 [Ref. 4], an internally stable lattice structure in 3-dimensional space consisting of **B** bars, **N** nodes and **S** supports, is stable if **B + S ≥ 3N**. For considerations about **S** see 1, 2 and 3. The redundancy of the system is:

$$R = B + S - 3N$$

✧ The similar considerations for plate structures:

Three supporting shear lines stabilize a plane plate.

Two plane plates linked by one shear-line are stabilized by five supports.

Three plane plates linked by three shear-lines - and any other 3-dimensional internally stable plate configurations - are stabilized by six supports.

As stated by Wester [Ref. 5], an internally stable plate structure in 3-dimensional space consisting of **SL** shear-lines, **P** plates and **S** supports, is stable if **SL + S ≥ 3P**. For considerations about S see 5, 6 and 7. The redundancy is

$$R = SL + S - 3P$$

✧ For combined plate and lattice structures:

As stated by Wester [Ref. 7], an internally stable combined plate and lattice structure in 3-dimensional space consisting of **B** bars, **SL** shear-lines, **BU** buffer-forces (i.e., forces acting between equally positioned bars and shear-lines), **N** nodes, **P** plates and **S** supports, is stable if

$$B + SL + BU + S \geq 3(N + P)$$

and the redundancy will become

$$R = B + SL + BU + S - 3(N + P)$$

2. The Genus of Polyhedra

The genus **g** is one of the parametrical axes in Burt's three-dimensional table. It is related to the connectivity **c**, which is the maximum number of loops that can be drawn on the surface, without dividing it into two parts. **g** is defined as the number of handles **h** of the polyhedron. **g = h = c – 1**.

A conventional single-connected polyhedron has no handles, hence **g = 0**. A torus has one handle, hence **g = 1**. Figure 3 shows polyhedra of different genus.

3. Plane Faceted Polyhedra - Genus Zero

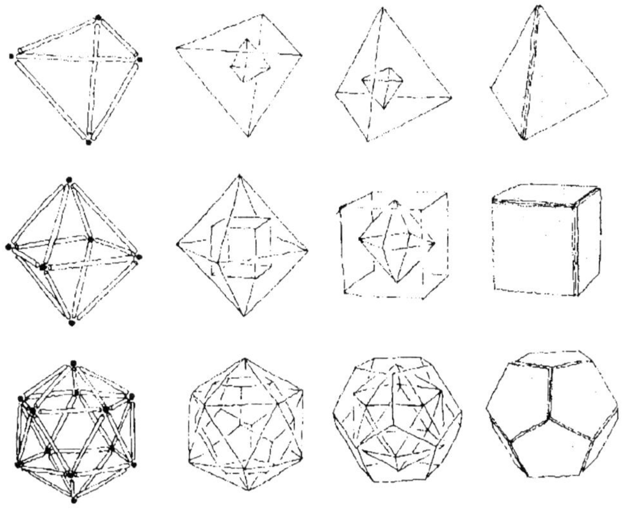

Fig. 1. The five archetypal Platonic plane faceted polyhedra arranged in dual pairs. The topological duality is extended to a structural duality, relating the stability behaviour of pure plate and pure lattice action. If one side of a dual pair is stable as a pure plate structure, then the other is stable as a pure lattice structure. The Platonic master-solid – the tetrahedron - is so basic that it is self-dual, hence stable as both a pure lattice and a pure plate structure. (Courtesy of Ola Wedebrunn)

This is the classic type of polyhedra, which follows Euler's Theorem in its basic version:

$$V + F - E = 2$$

Where the number of vertices is **V**, facets is **F** and edges is **E**. This case is structurally already described on the level of stability/topology [Refs. 5 & 7]. The summary of these investigations is that all plane-faceted Euler-polyhedra of this genus are internally stable with zero redundancy. As the minimum of number of supports **S** is 6, the previously stated stability equations will be:

✧ Pure lattice polyhedra: **B + 6 ≥ 3N** which implies the geometrical quality that all facets are trivalent. The redundancy is:

$$R = B - 3N + 6$$

✧ Pure plate polyhedra: **SL + 6 ≥ 3P** which implies the geometrical quality that all vertices are trivalent. The redundancy is:

$$R = SL - 3P + 6$$

✧ Combined plate and lattice polyhedra: $B + SL + BU + 6 \geq 3(N + P)$. The redundancy is $R = B + SL + BU + 6 - 3(N + P)$. If we use geometrical terms then B, SL and BU are all equal to E and therefore $R = 0$ for any arbitrary combined plate and lattice structure which follows Euler's theorem. This is valid for any combination of vertex and facet valences. The equations for pure plate and lattice polyhedra can be deduced from this, see [Ref. 7]. This stability behavior for genus zero polyhedra is probably the most important structural quality of all for polyhedra, as stability is – except for some rare configurations – inherent with the topology.

It is seen that if **B** is interchanged with **SL**, and **N** with **P**, the equations in 10 and 11 become interchangeable, an unchanged for 12. As this type of interchange of elements is following the rules for the classical geometrical duality, it turns out also to be the basis for a structural duality, see Fig. 1.

On the level of metric geometry/magnitude of forces the static relations are described in [Ref. 6] as follows: The geometrical part of the dual transformation is the polar reciprocation [Ref. 3] between vertices as nodes and planes as plates. It's characteristic is that the distance from Origin to the point and the are reciprocal and the normal vectors are located in the same line through the Origin.

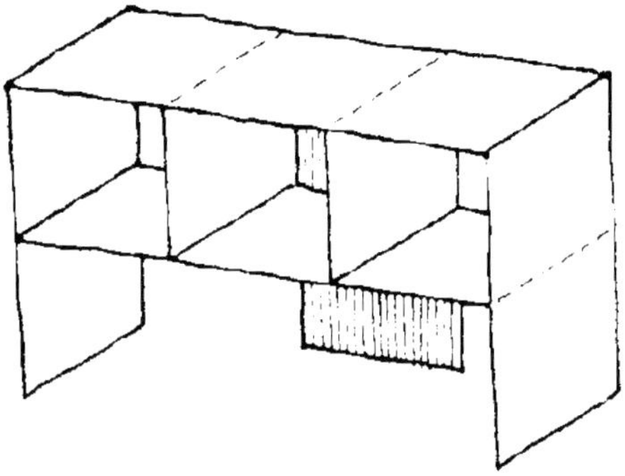

Fig. 2. Example of a non-polyhedral building-like structure, which utilizes plate action not only for stabilizing for horizontal loading, but also for carrying the two slabs. See later note on Buhelt, Nielsen & Staalby-76.

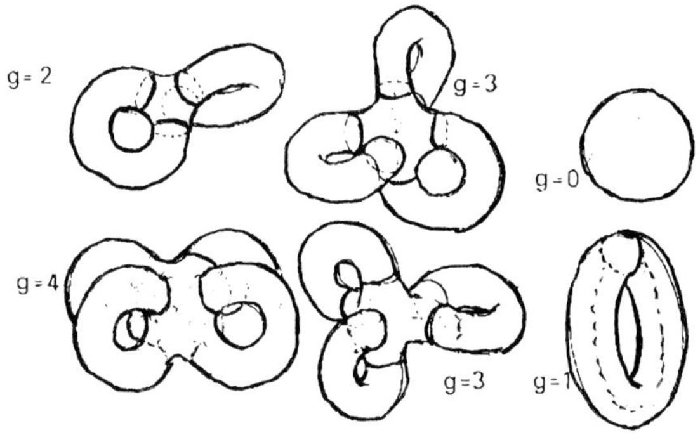

Fig. 3. Examples of polyhedra of different genus. It also indicates the way of determine the genus of units for infinite polyhedra. (Courtesy of Michael Burt)

The statical part of the dual transformation is that a force vector acting on a node is identical to a moment vector of the magnitude of the shear force times the distance to the Origin. This is based on the fact that a closed and uni-directed force vector polygon determines the static equilibrium of a node, while a closed and uni-directed moment vector polygon - based on the shear forces along the edges of a plate times their respective distances to the Origin - determines the static equilibrium of a plate. When these pairs of vector diagrams are identical, they form total equational identity between the two equilibrium systems.

This structural duality applies not only to all plane faceted polyhedra, but to all plane faceted structures in general (Fig. 2), and performs a method to do direct static analysis on any complicated plate structure by transforming it to its dual, do the analysis by one of the many lattice structure analysis software, and transform back the results. As the elastic properties also can be dual transformed [Ref. 6], then redundant structures may be analyzed through dual transformations.

4. Plane Faceted Polyhedra - Genus above Zero

For this case Euler's extended theorem is as follows:

$$V + F - E = 2(1\text{-}g)$$

Where the number of vertices is **V**, the number of facets is **F** and the number of edges is **E** while **g** is the genus. See Fig. 3.

As we still consider 3-dimensional finite structures, the basic stability equations are the same as for any genus number. This means that the stability equation for an arbitrary plane facetted polyhedron of genus **g**, is:

$$B + SL + BU + S \geq 3(N + P)$$

for a free floating polyhedron (**S** equal to 6)

$$B + SL + BU \geq 3(N + P - 2)$$

The redundancy is then

$$R = 3(E - V - F + 2) = 6g$$

This tells that the redundancy for combined plate and lattice polyhedral structure is only dependant of the genus, and is equal to six times the genus.

This means that a fully triangulated polyhedron of genus g has a redundancy of 6g as a pure lattice structure and therefore 6g bars may be removed from the structure without affecting the stability. In the same way, a fully trivalenced pure plate polyhedron of genus **g** has a redundancy of **6g**, and therefore **6g** shear-lines may be removed from the structure without affecting the stability.

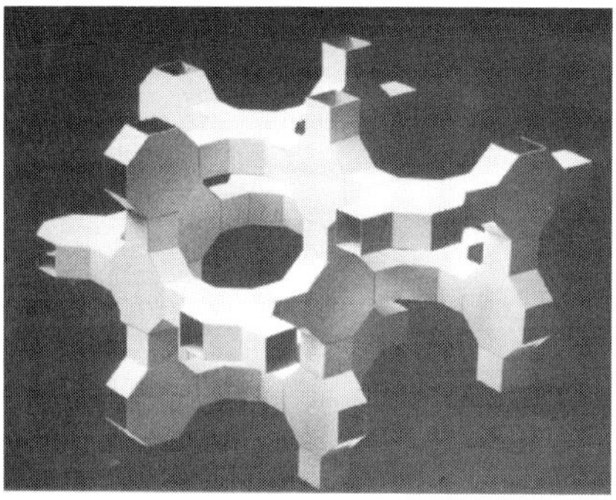

Fig. 4. Example of an infinite polyhedron of genus 2. It is stabilised as a combination of plate and lattice action. The redundancy is 6 for every internal unit, indicating that the structure might be structurally reduced without affecting the stability. (Courtesy of Michael Burt)

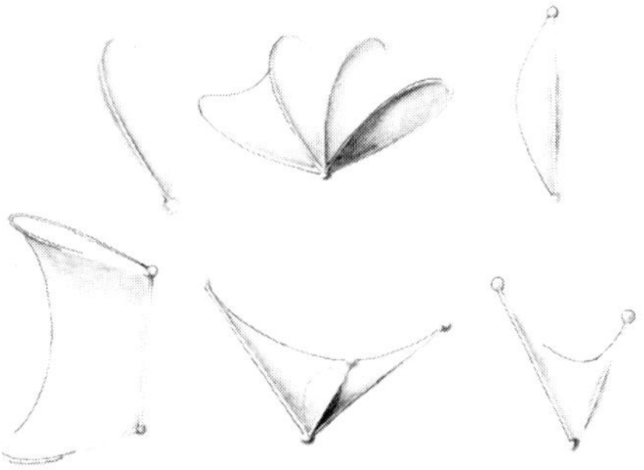

Fig. 5. Examples of floral polyhedra of genus zero. Many of these polyhedra are not imaginable with plane facets and straight edges, if they should enclose a volume. (Courtesy of Michael Burt)

5. Infinite Polyhedra

Infinite polyhedra are polyhedra, which are produced by sponge-like open polyhedral units. An infinite number of these units are then connected and closed along the open facets in a translational or/and rotational system without any limit, see Fig. 4. The unit is topologically described by the number of handles it will get (= **g**), if the missing facets on the unit are interconnected and closed with tubes, see Fig. 3.

As there are no support requirements for infinite polyhedra, the stability equation is

$$B + SL + BU \geq 3(N + P)$$

And therefore

$$R = 3(E - V - F) = 6(g\text{-}1)$$

Stability considerations are referred to the repeatable unit, which means that only periodic systems can be dealt with.

The number of bars and/or shear-lines that may be removed without affecting the stability situation is therefore **6(g-1)** per unit. The stability considerations for infinite polyhedra are useful for describing the basic structural qualities for large systems i.e.; systems where the effect of the border is small compared to the effect of the internal.

6. Floral Polyhedra

Floral polyhedra i.e., polyhedra with curved faces and edges, could be regarded as the general type of polyhedra, as the plane and the straight are special cases while the curved is the general case, see Fig. 5. These floral polyhedra are included in Burt's table simply because many can't be imagined with plane facets and they need to be there as they comply with Euler's theorem. A plane facetted polyhedron is a floral polyhedron where all the necessary curvature has been concentrated in the vertices. At the same time floral polyhedra increase the understanding of the polyhedral concept – and perform of course a tremendous challenge to bring them into a global structural framework which includes the framework described in the previous part of the paper.

A floral polyhedron consists of a number of zero-dimensional vertices, one-dimensional arbitrarily curved edges and two-dimensional arbitrarily doubly curved facets. Many of the polyhedra in Burt's table e.g., monogons, dihedrons, polydigons etc. can not be imagined as 3-dimensional volumes with straight and plane elements.

Not least because we know that undisturbed forces are acting along straight bar-lines, and shear fields are acting in plane plates, it could at the first glance seem quite impossible to describe any meaningful structural quality of floral polyhedra – and how could the concept of structural duality possibly be have any meaning? If plate and lattice action are defined by the type of force acting along the polyhedral edges i.e., axial forces for lattice structures and shear forces for plate structures, then the facet – curved or plane – acts as a stabilizing membrane between the edges, transferring forces which are either tangential (shear) or perpendicular (lattice) to the edge - all in equilibrium. If so, then we might describe the structural behavior of such polyhedra, see Fig. 6.

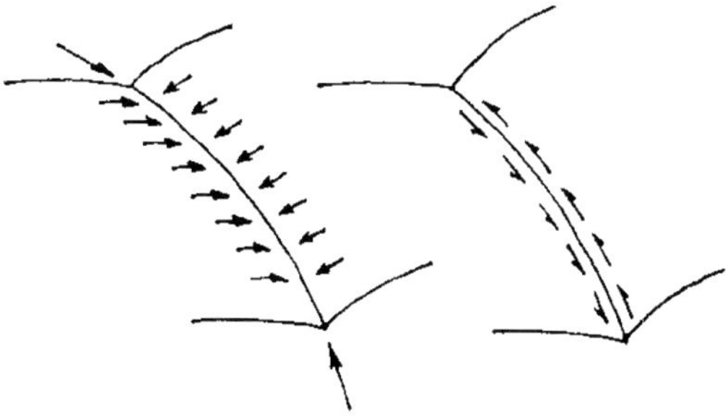

Fig. 6. We might define the structural type according to what kind of forces is transmitted from the curved facets to the curved edge. If the edge is in tension or compression and no shear is transferred, we talk of pure lattice action, while if it is only transferring shear forces, we talk about pure plate action. In all other cases we talk about combined action.

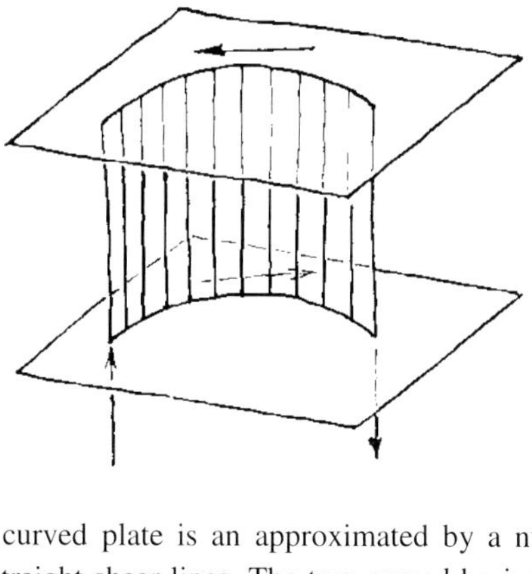

Fig. 7. A cylindrically curved plate is an approximated by a number of plane plates, intersecting in parallel straight shear-lines. The two curved horizontal shear-lines can be substituted with a straight shear-line with identical stabilising effect, using Bredt's equation.

This viewpoint is supported by what we already know about arbitrarily fine faceted i.e., curved polyhedra:

All closed plane faceted polyhedra – no matter how fine-faceted - are potentially stable, and the redundancy is only related to the genus.

- Any floral polyhedron may be regarded as consisting of many small plane facets and straight edges (as the discrete net in FEM), hence stable as a plane faceted pure lattice (triangles), pure plate (trivalent vertices) or combined structure with unchanged redundancy. Note that the structural action is unambiguously determined by the choice of faceting method.
- This also indicates that the stability of a floral polyhedron may be described in the same terms as for the plane faceted polyhedra.

Also we know from [Ref. 1] that some curved plates in combination with other plates in a building like structure can act exactly the same way as a plane plate i.e., it can be substituted structurally by a virtual plane plate which is located according to some simple static considerations. The considered plates are curved with straight generators, such as cylindrical, conical and HP-shapes.

The curved plate e.g., a vertical cylindrical wall is placed between two horizontal plates (slabs) in a building, where the lower plate is stabilized for all horizontal forces and the two lower ends of the curved plate stands on vertical columns. The task is then to find the horizontal straight line in the upper plate where the curved plate resists external load. This line turns out to be parallel to the line between the upper end points of the cylindrical wall, and the distance to this line - the eccentricity **e** of the force - is found by Bredt's equation, which tells that **e = 2A/a**, where **A** is the area between the trace of the curved plate and its chord – and **a** is the distance between the same two end points, see Fig. 7.

The polyhedral situation is different from the above consideration as intersecting facets creates edges and vertices, altogether forming closed configurations. If the configuration with the previously mentioned cylindrical plate is changed so it is inserted and connected as a diaphragm into a close-fit rectangular tube, which is open in both ends, it forms an unstable polyhedron which miss one facet to be closed. This configuration is applicable for investigating the efficiency of the diaphragm as a stabilizing element.

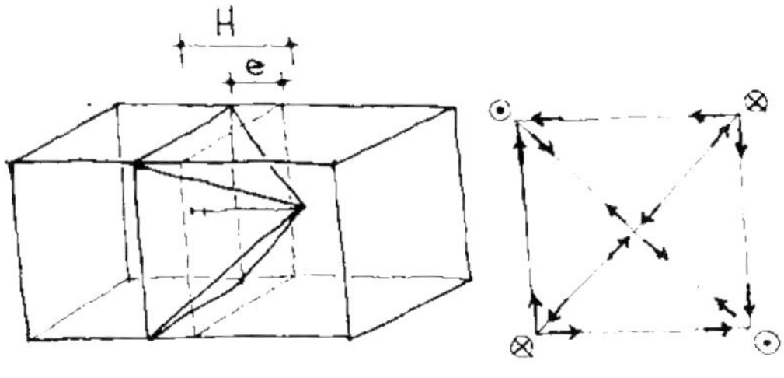

Fig. 8. Left is shown the pyramid with open rectangular bottom; inserted in a rectangular tube, which is open in both ends. The height of the tube is H and the distance from the base of the pyramid to the plane virtual plate with the identical stabilizing effect is called the eccentricity e. To the right is the diagram of forces is equilibrium for the system. The forces perpendicular to the plane of the paper are creating the moment that moves the applied shear-force the distance e, following Bredt's equation.

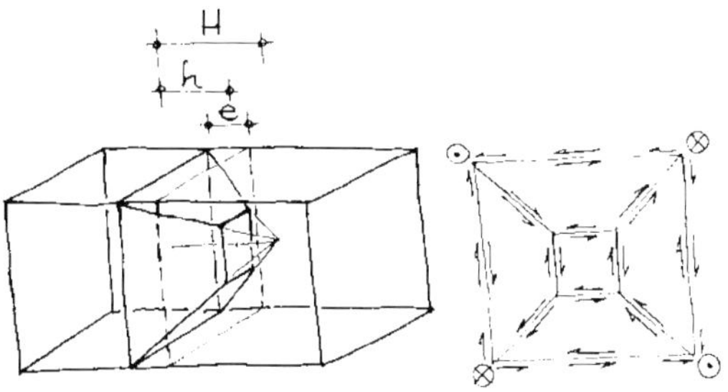

Fig. 9. If the pyramid is truncated, then the system is changed to plate action. The diagram of forces in equilibrium is shown to the right. In this case e is not following Bredt's equation, as the shear stresses transferred are not of the same intensity.

The eccentricity **e** creating the moment, which is needed for stabilizing the configuration is therefore acting on four side sides of the tube instead of the two horizontal plates. This leads - according to Bredt's equation - to a position of the virtual plane plate with an eccentricity **e = A/a**.

A method to determine the structural action of a plate (or manifold) with positive Gaussian curvature is to approximate its shape with plane facets, and perform the necessary statical considerations. A simple approximation is a four-sided pyramid as a half octahedron with an open bottom.

The half octahedron Fig. 8 is an unstable lattice net of four closed triangles and one open rectangle with variable height **H** - substituting a positively doubly curved plate - is inserted as a diaphragm in the rectangular tube. A physical model shows clearly that stability is achieved for certain positions of forces applied to the planes of the tube. The moment produced in the planes of the tube moves the virtual plane plate to a distance **e = H/2** from the shear line at the bottom edge of the pyramid. This case follows Bredt's equation as the shear-line has the length of **a**, the area **A = aH/2**, hence **e = A/a = H/2**.

A better approximation to the elliptically curved plate is the truncated pyramid Fig. 9. If the height of the full pyramid is **H** and the height of the truncated pyramid is h, then the statical consideration leads to the nice equation:

$$\mathbf{e} = \frac{1}{\frac{1}{\mathbf{H}} + \frac{1}{\mathbf{h}}}$$

If $\mathbf{H} \rightarrow \mathbf{0}$ or if $\mathbf{h} \rightarrow \mathbf{0}$ then $\mathbf{e} \rightarrow \mathbf{0}$, which is correct as this corresponds to a plane plate. If $\mathbf{h} \rightarrow \mathbf{H}$ then $\mathbf{e} \rightarrow \mathbf{H/2}$, which is also correct as this corresponds to the previously investigated pyramid. Because the forces in the edges of the truncated pyramid are not constant, then **e** does not follow Bredt's equation.

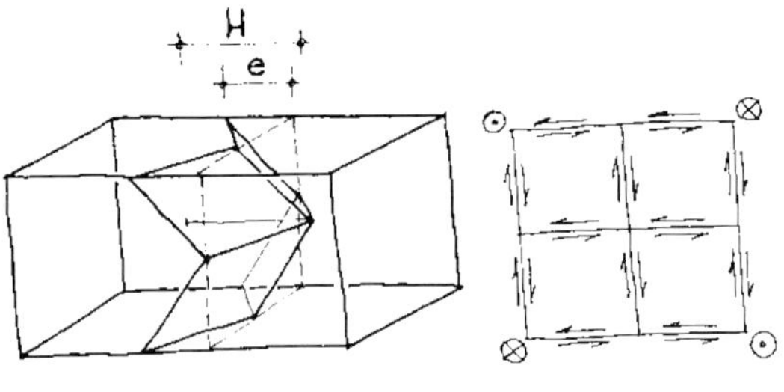

Fig. 10. If the pyramid is rotated 45°, then the system appears like this. The equilibrium is achieved as pure plate action. The eccentricity e is found to follow Bredt's equation.

If the half lattice octahedron is rotated 45° as Fig. 10, then the configuration is changed to plate action, and the eccentricity to the stable plane virtual plate is found to **e = H/2**, hence still complying with Bredt's equation. The four edges meeting in the top of the pyramid can obviously not transfer axial forces; even the valence is bigger than what is required for pure plate action.

If the top of the pyramid on Fig. 10 is truncated, then the eccentricity is not changed at all. This means that the rectangular plate positioned in the truncation is not structurally active, and may therefore be omitted.

7. Conclusion

My recent work on doubly curved plates, which is initiated by Burt's inspiring work on geometry and topology, and based on Jørgen Nielsen's work on the static of cylindrically curved plates, seems to indicate, that it is absolutely possible to deal with the structural behavior of floral polyhedra in a meaningful way. We can describe the stability behavior of curved plate or lattice or combined plate and lattice membrane diaphragms and we can find the position of a virtual plane plate with identical stabilizing qualities as the curved one. The investigations of curved plates in the present paper should be extended to saddle shapes and to intersecting curved shapes.

It is still not investigated exactly how the structural duality apply to floral polyhedra, but it is beyond any doubt that there is a dual relation of the structural content, and that the concept will include my earlier work on the structural duality behavior of plane faceted polyhedra.

References and Notes

1. Buhelt M., Nielsen J., Staalby J. 1976. Skivebygningers Stabilitet 1, Konstruktionsprincipper (Plate Stabilised Buildings: Structural Principles). The Danish Building Research Institute, recommendation no. 82 (in Danish). This report describes the structural action of plates as a main load-carrying element in buildings. It also deals with curved plates and how to find the position of the virtual plane plate that has exactly the same stabilizing effect in the building structure as the

curved plate, by using Bredt's equation. It is probably the first time that structural performance of curved plates has been considered. The report is basically inspired by student's works in courses in plate structures (with the author as co-teacher) at the Royal Danish Academy, School of Architecture.

2. Burt M. The Periodic Table of the Polyhedra Universe. Technion Publication, Haifa, Israel 1996.
3. Cundy H.M., Rollet A.P. Mathematical Models Third edition. pp. 78-79. Tarquin Publications. 1981. The geometrical content of polar reciprocation is thoroughly explained in Cundy and Rollet's classical book on the geometry of polyhedra.
4. Moebius A.F., Lehrbuch der Statik. Georg Joachim Göschen, Leipzig 1837. (also inventor of the non-orientable strip) wrote this remarkable book on statics where you will find several considerations, which seems to have been almost forgotten, and reinvented by others. Some of them are still on the research agenda, e.g. the structural behavior of polyhedra, and infinitesimal movability of certain structural configurations.
5. Wester T. Structural Order in space - the Plate Lattice Dualism. Royal Danish Academy of Fine Arts, 1983. The results were originally stated in the author's research report in Danish in 1976. According to correspondence with the Canadian mathematician Dr. Walter Whiteley, it was found independently by him at approximately the same time, see Whitely W. Rigidity and Polarity I: Statics of Sheet works. Geom. Dedicata 22, pp. 329-362, 1987.
6. Wester T. The Plate-Lattice Dualism. Paper in T.T. Lan & Y. Zhilihan (ed) Proceedings of the international colloquium in Beijing on Space Structures for Sports Buildings, pp. 321-328. Elsevier Applied Science, London, 1987.This paper, presented at the IASS Colloquium in Beijing, deals with the duality on the level of metric geometry and magnitude of forces, including transformation of elastic properties.
7. Wester T. The Structural Behavior of Arbitrarily Plane-Faceted Spatial nets. Paper in T. Wester, S.J. Medwadowsky, I. Mogensen (eds) proceedings for the IASS Symposium on Spatial Structures at the Turn of the Millennium, Royal Danish Academy of Fine

Arts, 1991, Vol. 2, pp. 119-124. This paper, presented at the IASS Symposium in Copenhagen, deduces the necessary stability equation for any plane facetted conventional polyhedron as a combination of plate and lattice action. For the structural analysis of plane facetted polyhedra of higher genus, see also Wester, T. & Burt, M. The Basic Structural Content of the Periodic Table of the Polyhedral Universe. Proceedings of the IASS Symposium in Singapore 1997, pp. 869-876. This paper is the first attempt to merge the geometric and the structural approach, and The Structural Morphology of the Polyhedral Universe: Preliminary Considerations. Proceedings of the international Conference "Engineering a New Architecture" pp. 197-206, School of Architecture, Aarhus 1998.

Courtesy of Michael Burt

CHAPTER 4

POLYHEDROIDS

Pieter Huybers

Assoc. Professor, Delft Univ. of Technology, Fac. CiTG, 2628CN, Stevinweg 1, Delft, The Netherlands

The title of this paper refers to the group of figures, that are related to polyhedra, that are - so to say - 'polyhedron-like'. They are derived from the Platonic (or regular) and Archimedean (or semi-regular) polyhedra, that are composed of regular polygons. They can be quite different from the original figures, that they are based on. This paper pays attention to a family of forms related to these so-called 'uniform' polyhedra. Possibilities are shown to mitigate the rigidity of the polyhedral geometry and to make them more suitable for application in building construction.

1. Definition of Polyhedra

A common definition of a polyhedron is [Ref. 1]:

(i) It is covered by a closed pattern of plane, regular polygons with 3, 4, 5, 6, 8 or 10 edges.

(ii) All vertices of a polyhedron lie on one circumscribed sphere.

(iii) All these vertices are identical. In a particular polyhedron the polygons are grouped around each vertex in the same number, kind and order of sequence.

(iv) The polygons meet in pairs at a common edge.

(v) The dihedral angle at an edge is convex. In other words: the sum of the polygon face angles that meet at a vertex is always smaller than 360° (see Table 2).

uncomfortably difficult scientific names, but they can also be used in a more or less "administrative sense". If one utilizes computer programs for the calculation of their geometry or for their visual presentation, it is often necessary to indicate them by a unique number. That is why they are numbered here in a certain order of sequence that is dictated by the following consecutive criteria:

1. Number of faces
2. Number of edges
3. Radius of the circumscribed sphere

If only criteria 1 and 2 were applied, the truncated dodecahedron and the truncated icosahedron would have obtained the same number. The left- and right-handed snubs have the same identification numbers, because they are geometrically identical, although they have different co-ordinates. They are sometimes called 'chiral' [Ref. 14]. The two 'enantiomorphic' versions can be distinguished by the addition of an L or an R to the index number. The polygons are facets of the circumscribed

Table 3. Review of the numbers in which the polygons occur in the different polyhedra and their Volume, Area and Compactness, expressed in unit edge length.

P	3	4	5	6	8	10	Volume	Area	Compactness
1	4	-	-	-	-	-	0.11785	1.73205	0.67113
2	-	6	-	-	-	-	1.00000	6.00000	0.80599
3	8	-	-	-	-	-	0.47140	3.46410	0.84558
4	-	-	12	-	-	-	7.66311	20.64572	0.91045
5	20	-	-	-	-	-	2.18169	8.66025	0.93932
6	4	-	-	4	-	-	2.71057	12.12435	0.77541
7	8	6	-	-	-	-	2.35702	9.46410	0.90499
8	-	6	-	8	-	-	11.31370	26.78460	0.90991
9	8	-	-	-	6	-	13.59966	32.43466	0.84949
10	8	18	-	-	-	-	8.71404	21.46410	0.95407
11	-	12	-	8	6	-	41.79898	61.75517	0.94316
12	20	-	12	-	-	-	13.83552	29.30598	0.95102
13	-	-	12	20	-	-	55.28773	72.60725	0.96662
14	20	-	-	-	-	12	85.03966	100.99076	0.92601
15	32	6	-	-	-	-	7.88947	19.85640	0.96519
16	20	30	12	-	-	-	41.61532	59.30598	0.97923
17	-	30	-	20	-	12	206.80339	174.29203	0.97031
18	80	-	12	-	-	-	37.61664	55.28674	0.98201

sphere that can be thought to pass through the vertices. The volume of this sphere is therefore larger than that of the corresponding polyhedron. This is also the case for the area of their envelopes. The closer these two values are, the better is the approximation of the sphere that is reached by a particular polyhedron. The closeness of this approximation can be expressed in a value that is called: the Compactness of a polyhedron.

Compactness C_p = Quotient of the area of a sphere with the same volume as the polyhedron with the index P, divided by the surface area of this polyhedron.

This value is given in the equation:

$$\text{Compactness } C_P = \frac{\sqrt[3]{36\pi * Vol_P{}^2}}{Area_P} \tag{1}$$

The dual or reciprocal versions of the polyhedra can be indicated by the same number as their mates, but this time, they are preceded by the capital R.

4. Close-Packings

Some of the polyhedra lend themselves to being put together in tight packed formations (Fig. 4). In this way quite complex forms can be realized. It is obvious hat cubes and rectangular prisms can be stacked most densely, but many of the other polyhedra can also be packed in certain combinations. Critchlow [Ref. 1] gives an extensive review of all theoretical possibilities.

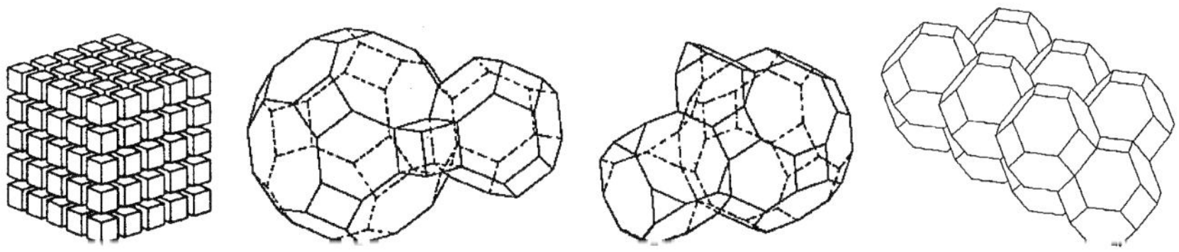

Fig. 4. Examples of tight polyhedral packing.

5. Prisms and Antiprisms

Other solids that also respond the previous definition of a polyhedron are the prisms and the anti-prisms. Prisms have two parallel polygons like the lid and the bottom of a box and square side-faces; anti-prisms are like the prisms but have one of the polygons slightly rotated so as to turn the side-faces into triangles.

(5)

(6)

Figs. 5 and 6. Models of prisms and anti-prisms.

The simplest forms are the prismatic shapes. They fit usually well together and they allow the formation of many variations of close-packing. If a number of anti-prisms is put together according to their polygonal faces, a geometry is obtained of which the outer mantle has the appearance of a cylindrical, concertina-like folded plane. [Ref. 6]

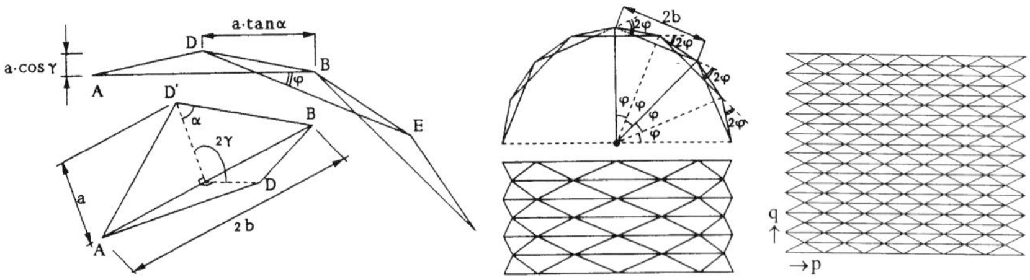

Fig. 7. Variables that define the shape of antiprismatic forms.

These forms can be described with the help of only few parameters, a combination of 3 angles: α, β and γ. The element in Fig. 7A represents 2 adjacent isosceles triangles.

α = half the top angle of the isosceles triangle ABC with height a and base length 2b.
γ = half the dihedral angle between the 2 triangles along the basis.
φ_n = half the angle under which this basis with the length 2b is seen from the cylinder axis. = π/n.

The relation of these angles α, γ and φ_n [Ref. 3]:

$$\tan \alpha = \cos \gamma \ \text{cotan} (\varphi n/2) \tag{2}$$

These three parameters define together with the base length (or scale factor 2b) the shape and the dimensions of a section in such a structure. This provides an interesting tool to describe any anti-prismatic configuration. Two additional data must be given: the number of elements in transverse direction (p) and that in length direction (q).

6. Augmentation

Upon the regular faces of the polyhedra other figures can be placed that have the same basis as the respective polygon. In this way polyhedra can be 'pyramidized'. This means that shallow pyramids are put on top of the polyhedral faces, having their apexes on the circumscribed sphere of the whole figure. This can be considered as the first frequency subdivision of spheres. In 1582 Simon Stevin introduced the notion of 'augmentation' by adding pyramids, consisting of triangles and having a triangle, a square or a pentagon for its base, to the 5 regular polyhedra [Ref. 4]. Recently, in 1990 D.G. Emmerich extended this idea to the semi-regular polyhedra (Fig. 8). He suggested to use pyramids of 3-, 4-, 5-, 6-, 8- or 10-sided base, composed of regular polygons, and he found that 102 different combinations can be made. He calls these: composite polyhedra [Ref. 5].

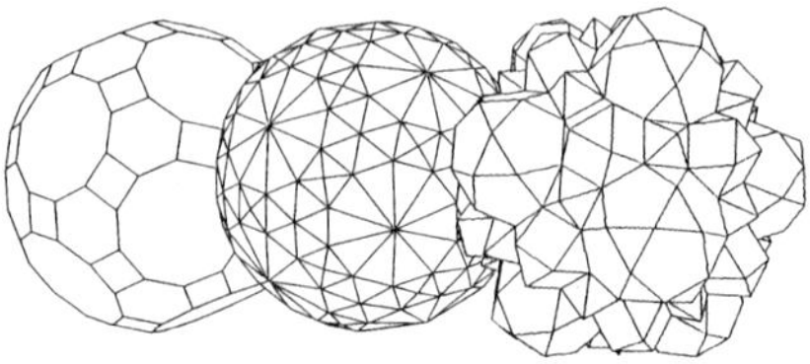

Fig. 8. Augmented P17.

(9)

(10)

Figs. 9 and 10. Models of zonohedral additions.

7. Sphere Subdivisions

For the further subdivision of spherical surfaces generally the icosahedron – and in some cases the tetrahedron or the octahedron - are used as the starting point, because they consist of equilateral triangles that can be easily covered with a suitable pattern that is subsequently projected upon a sphere. This leads to economical kinds of subdivision up to high frequencies and with small numbers of different member lengths [Ref. 7].

Fig. 11. Models of various dome subdivision methods.

Fig. 12. Great circle subdivisions [Ref. 8].

All other regular and semi-regular solids, and even their reciprocals as well as prisms and anti-prisms can be used similarly [Ref. 12]. The polygonal faces are first subdivided and then made spherical.

8. Sphere Deformation

The spherical co-ordinates can be written in a general form, so that the shape of the sphere may be modified. This leads to interesting new shapes that all have the same basis but are governed by different parameters (Fig. 13A). According to H. Kenner [Ref. 9] the equation of the sphere can be transformed into a set of two expressions, describing it in a more general way:

$$R_1 = E_1 / (E_1^{n1} \sin^{n1}\varphi + \cos^{n1}\varphi)^{1/n1} \tag{3}$$

$$R_2 = R_1 E_2 / (E_2^{n2} \sin^{n2}\theta + R_1^{n2} \cos^{n2}\theta)^{1/n2} \tag{4}$$

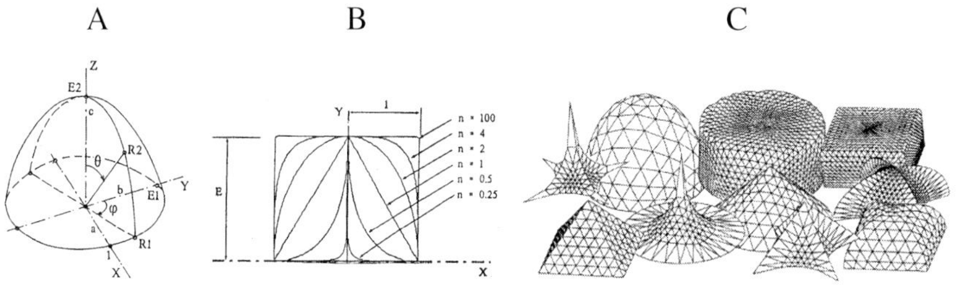

Fig. 13. Form variation of domes by use of different variables.

n1 and n2 are the exponents of the horizontal and vertical ellipse and E1 and E2 the ratios of their axes. The shape of the sphere can be altered in many ways, leading to a number of transformations. The curvature is a pure ellipse if $n = 2$, but if n is raised a form is found, which approximates the circumscribed rectangle. If n is decreased, the curvature flattens until $n = 1$ and the ellipse then has the form of a pure rhombus with straight sides, connecting the maxima on the co-ordinate axes. For $n < 1$ the curvature becomes concave and obtains a shape, reminiscing a hyperbola. For n = 0 the figure coincides completely with the X-, and Y-axes. By changing the value of both the horizontal and the vertical exponent the visual appearance of a hemispherical shape can be altered considerably. [Refs. 9 and 10]

9. Polyhedra in Building

The role that polyhedra can play in the form-giving of buildings is very important, although this is not always fully acknowledged. Some possible or actual applications are referred to here briefly.

9.1. *Cubic and Prismatic Shapes*

Most of our present-day architectural forms are prismatic with the cube as the most generally adopted representing shape. Prisms are used in a vertical or in a horizontal position, in pure form or in distorted versions. This family of figures is therefore of utmost importance for building.

(14)

(15)

Figs. 14 and 15. Models of a space frames made of square plates or of identical struts.

9.2. *Solitary Polyhedra*

Architecture can become more versatile and interesting with macro-forms, derived from one of the more complex polyhedra or from their reciprocal (dual) forms, although this has not often been done. Packing of augmented polyhedra form are sometimes interesting alternatives for the traditional building shapes.

(16)

(17)

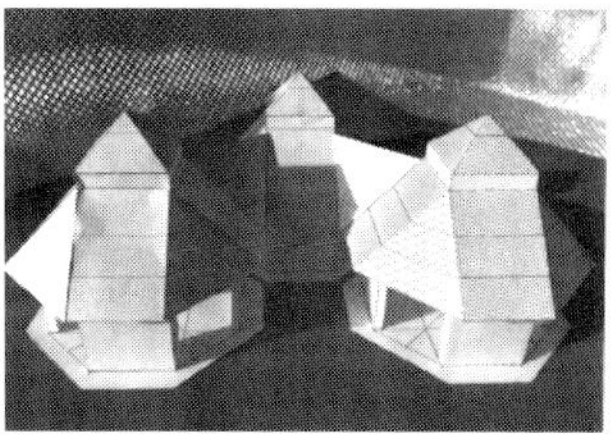

(18)

Figs. 16, 17 and 18. Models of houses based on P10, truncated R7 and P10 (built in Mali).

9.3. *Combinations*

Close-packing is also suitable as the basic configuration for space frames, because of their great uniformity. If these frames are based on tetrahedra or octahedra, all members are identical and meet at specific angles. Many of such structures have been built in the recent past and this has become a very important field of application. The members usually meet at joints having a spherical or a polyhedral form (Fig. 15).

9.4. *Domes*

R.B. Fuller re-discovered the geodesic dome principle. This has proven to be of great importance for the developments in this field. Many domes have been built during the last decades, up to very large spans. A new group of materials with promising potential has been called after him, which has molecules that basically consist of 60 atoms, placed at the corners of a truncated icosahedron or P13 (See Fig. 20).

10. 3D-Slide Presentation

The author has shown during the conference a few applications with the help of a 3-D colour slide presentation. Two pictures with different orientation of the light waves are projected simultaneously on one screen. The screen must have a metal surface which maintains these two ways of polarization, so that the two pictures can be observed with polaroïd spectacles that disentangle them again into a left and a right image. These pictures are made either analogously, which means: with normal photo cameras and with the two pictures token at a certain parallax. The same technique can be used for digital slides, were the pairs of pictures are made by computer and subsequently written directly onto positive film. This technique allows colored pictures to be shown in a really three-dimensional way and gives thus a true impression of the spatial properties of the object [Ref. 15].

Fig. 19. Pair of stereoscopic pictures.

Fig. 20. Model of small 'Fullerene'.

Fig. 21. Review of a great number of polyhedron based models.

References

1. Critchlow, K., Order in Space, Thames and Hudson, London, 1969.
2. Huybers, P., The geometry of uniform polyhedra, Architectural Science Review, Vol. 23, no. 2, June 1980, p. 36-50.
3. Huybers, P., Polyhedra and their reciprocals, Proc. IASS Conference on the Conceptual Design of Structures, 7-11 October, 1996, Stuttgart, 254-261.
4. Struik, D.J., The principle works of Simon Stevin, Vol. II, Swets & Seitlinger, Amsterdam, 1958.
5. Emmerich, D.G., Composite polyhedra. Int. Journal of Space Structures, 5, 1990, p. 281-296.
6. Huybers, P. and G. van der Ende, Prisms and antiprisms, Proc. Int. IASS Conf. on Spatial, Lattice and Tension Structures, Atlanta, 24-28 April 1994, p. 142-151.

7. Clinton, J.D., Advanced structural geometry studies, NASA Contract Report CR. 1735.
8. Wenninger, M., 1979, Spherical models, Cambridge University Press, USA.
9. Kenner, H., Geodesic math and how to use it, University of California Press, London, 1976.
10. Huybers, P., About the manipulation of spheroidal forms, Int. IASS Conf. Innovative Large Span Structures, Toronto, 12-17 July, 1992, Vol. 2, p. 55-66.
11. Huybers, P., Prismoidal structures, The Mouchel Centenary Conf. on Innovation in Civil & Structural Engineering, Cambridge, 19-21 August 1997, p. 79-C88.
12. Huybers, P. and G. van der Ende, Polyhedral sphere subdivisions, Int. IASS Conf. on Spatial Structures: Heritage, Present and Future, Milan 5-9 June, 1995, p. 189-198.
13. Huybers, P., The polyhedral world, In: 'Beyond the Cube: The Architecture of Space frames and Polyhedra', J.F. Gabriel Ed., John Wiley and Sons, Inc., New York, 1997, p. 243-279.
14. Huybers, P., The chiral Polyhedra, IASS Journal, Vol. 40, Nr. 2, August 1999, p. 133-143.
15. Ferwerda, J.G., The world of 3-D, a practical guide to stereo photography, 3-D Book Productions, Borger, 1987.

CHAPTER 5

NOVATIONAL TRANSFORMATIONS

H. Nooshin[1], F. G. A. Albermani[2] and P. L. Disney[1]

[1]*Space Structures Research Centre, Department of Civil Engineering University of Surrey, Guildford, Surrey GU2 5XH, UK*
[2]*Department of Civil Engineering, The University of Queensland Brisbane, Queensland 4072, Australia*

The term 'novational transformation' is used to refer to a particular kind of geometric transformation that allows the shape of a configuration to be modified by specifying one or more 'movements'. Novational transformations are of great value in configuration processing, in particular, in relation to shape generation for space structures. The objective of the present paper is to explain the general idea of a novational transformation and to describe the details of the novational transformations that are available for use in the programming language Formian. The material is presented in terms of a number of examples.

1. Introduction

'Configuration processing' is concerned with computer aided creation and manipulation of configurations and the programming language 'Formian' provides a suitable medium for configuration processing [Ref. 1]. Configuration processing activities are performed through a variety of conceptual tools. The tools include a number of families of transformations that allow the body of a configuration to be deformed in various ways. One such family of transformations is discussed in the present paper.

Figs. 2(a) and 2(b). Thus, the position of a node remains unchanged unless it has a directly specified translocation.

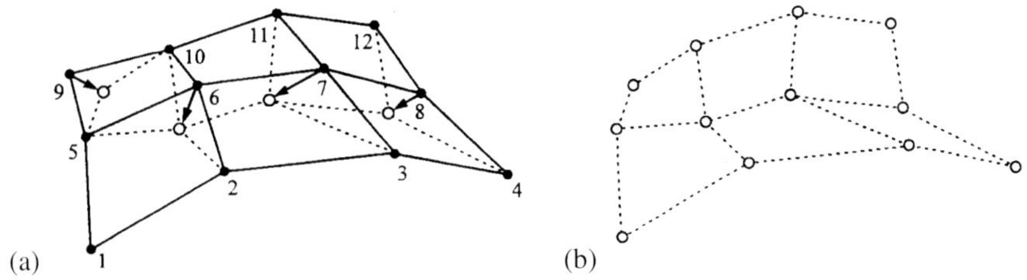

Fig. 2. An example of a sharp novation.

5. Exponential Decay Novations

Consider the planar configuration shown in Fig. 3 and let it be required to subject the configuration to an ED novation with a single specified translocation. In general, a specified translocation may or may not be at a nodal position. In the case illustrated in Fig. 3, the specified translocation is at a non-nodal position. To apply the ED novation, all the nodes of the configuration are considered one after the other starting from the first node. For a typical node j, the coordinates in the UI and U2 directions are modified as follows:

The kth coordinate of node j is modified by adding to it the quantity

$$T_k = T_{sk}(10^{-Cdj})$$

Where

k is either 1 or 2, indicating the 1st or 2nd coordinate direction,

T_{sk} is the kth component of the specified translocation, that is, T_{S1} or T_{S2} as shown in Fig. 3, where the subscript s indicates that T_{S1} and T_{S2} are components of the 'specified' translocation.

dj is the 'relative' distance of node j from the position of the specified translocation and is given by the ratio Dj/D, where,

D is the length of the diagonal of the 'box frame' of the configuration, as shown in Fig. 3, where, the box frame of a configuration is defined as the smallest rectangle (or rectangular solid or hyper-rectangular solid) that contains the configuration and whose edges are parallel to the coordinate axes,

Dj is the distance of the node j from the position of the specified translocation, as shown in Fig. 3, and

C is referred as the 'control parameter'.

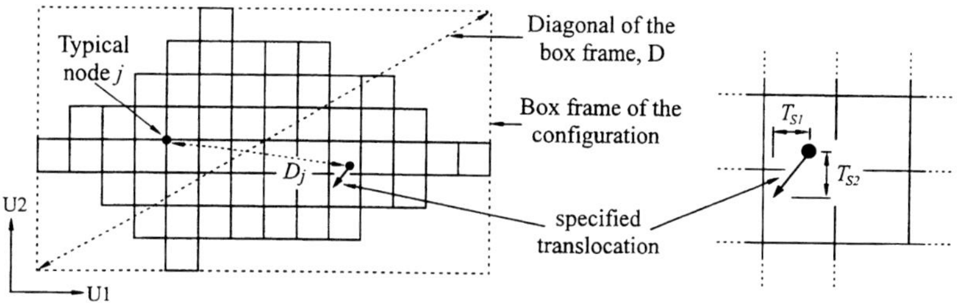

Fig. 3. A planar configuration with a specified translocation.

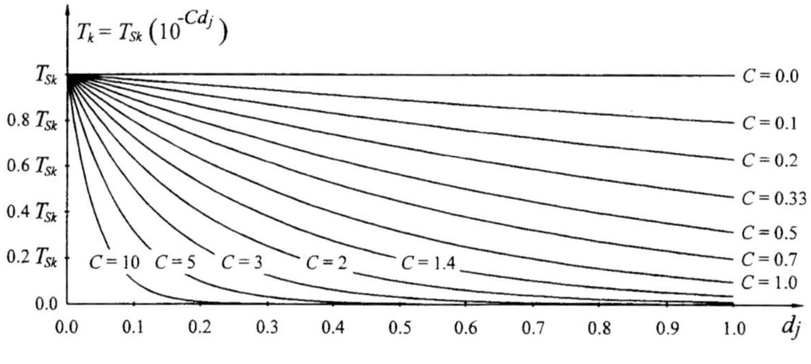

Fig. 4. Exponential decay curves.

The quantity $T_k = T_{sk}(10^{-Cdj})$ is plotted against dj, in the range dj = 0 to dj = 1, for different values of C in Fig. 4. The purpose of this figure is to illustrate how a component of specified translocation (that is, T_{Sk}) affects the coordinates of a typical node j. When C > 0, then the proportion of T_{Sk} that is transferred to node j (that is, T_k) varies between zero and unity, as shown in Fig. 4.

The curves in Fig. 4 relate to different values of C. It is seen that the proportion of T_{Sk} that is transferred to node j always decreases with increasing dj. However, the higher the value of C, the more rapid is the 'decay' of the curve. That is, the value of C is a measure of the rate of decay of the curve. In other words, the higher the value of C, the less the effect of T_{Sk} will be on the coordinates of the nodes of the configuration.

An example of the application of the ED novation is shown in Fig. 5(a). The planar configuration shown by full lines in Fig. 5(a) has 8 line elements and 6 nodes. There is only one specified translocation which is indicated by an arrow at node 5. The components of this translocation in the first and second directions are $T_{S1} = 2$ and $T_{S2} = 1$, respectively. The configuration is subjected to an ED novation with C = 1. The resulting configuration is shown by dotted lines, with the numerical values of the nodal coordinates and translocation components given in Fig. 5(b). The values given for node 5 in the last two columns of the table in Fig. 5(b) are the components of the 'specified' translocation but the values for the other nodes in these columns are the components of 'conformity' translocations.

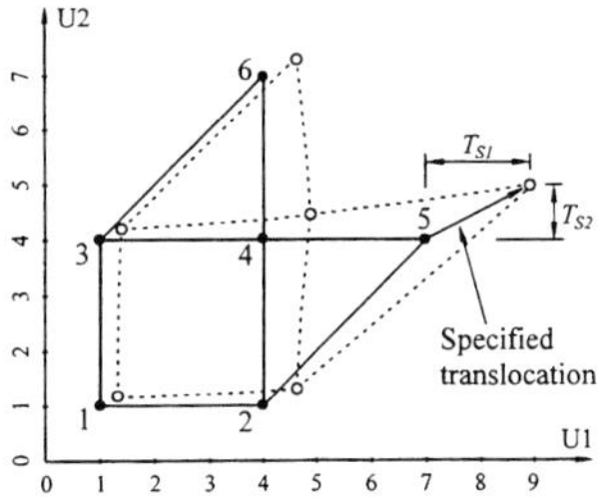

(a) Control parameter C is 1

Node	U1	U2	Translocation	
			T_1	T_2
1	1	1	0.32394	0.16197
2	4	1	0.63246	0.31623
3	1	4	0.39258	0.19629
4	4	4	0.88608	0.44304
5	7	4	**2**	**1**
6	4	7	0.63246	0.31623

(b)

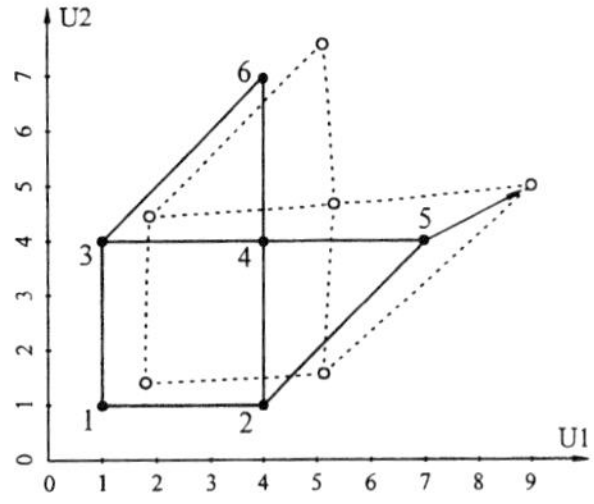

(c) Control parameter C is 0.5

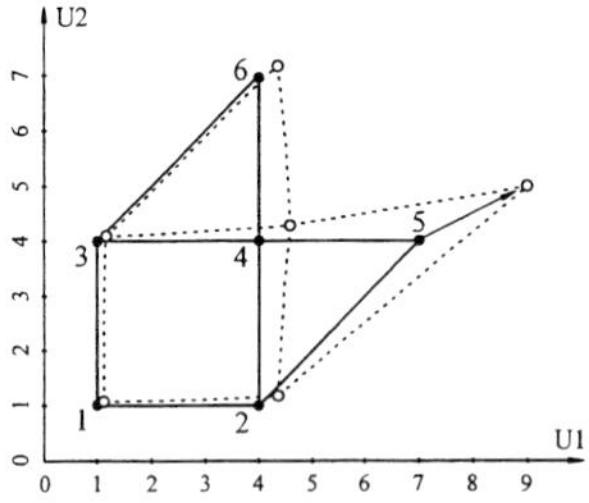

(d) Control parameter C is 1.5

Fig. 5. Example of ED novations with a single specified translocation.

To illustrate the effect of the control parameter C, the configuration of Fig. 5(a) is subjected to ED novations with another two values of C and the results are shown in Figs. 5(c) and 5(d). The value of the control parameter C in Fig. 5(c) is equal to 0.5. The comparison of Figs. 5(a) and 5(c) indicates that the conformity translocations in Fig. 5(c) are larger than the corresponding ones in Fig. 5(a). The reverse is true for the example shown in Fig. 5(d) where the control parameter C is equal to 1.5.

6. Multiple Specifications

Now suppose that the configuration of Fig. 5(a) has two specified translocations, as shown in Fig. 6(a). In this case, in addition to node 5 that has the same specified translocation as before, node 6 has a specified translocation whose components in the first and second directions are 1 and -1, respectively. In applying the ED novation in this situation, one may be tempted to think that the problem can be handled by a 'double application' of the procedure used in the example of Fig. 5. That is, by combining the results of the separate applications of the procedure for nodes 5 and 6 in Fig. 6(a). However, this strategy will fail to produce the desired effects. The problem is due to the 'interaction' between the specified translocations at nodes 5 and 6. Thus, if the 'double application' strategy is followed, the final translocations at nodes 5 and 6 will be different from the specified ones.

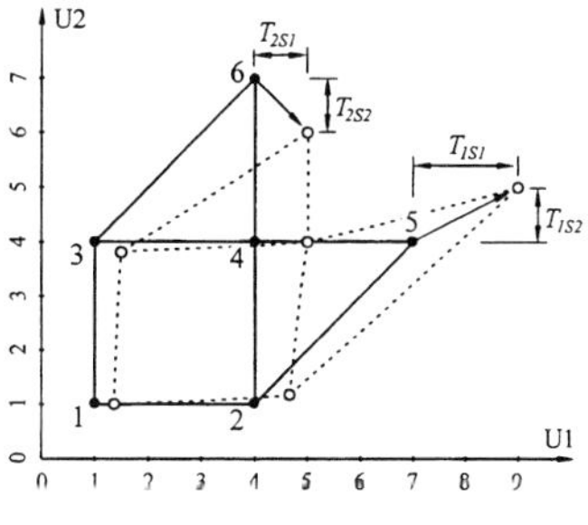

(a) Control parameter C is 1

Node	U1	U2	Translocation	
			T1	*T2*
1	1	1	0.36917	0
2	4	1	0.67178	0.17541
3	1	4	0.49637	-0.17541
4	4	4	1.00980	0
5	7	4	2	1
6	4	7	1	-1

(b)

Fig. 6. Example of an ED novation with two specified translocations.

The question is: What components of translocation should be imposed at nodes 5 and 6 so that the combined effects produce the correct values of the specified components of translocation? To answer the question, let the components of translocation that should be imposed at nodes 5 and 6 be denoted by T_{1W1}, T_{1W2}, T_{2W1}, and T_{2W2}. These are referred to as 'working translocation components' and correspond, respectively, to the specified translocation components T_{1S1}, T_{1S2}, T_{2S1}, and T_{2S2}, shown in Fig. 6(a). In general, a component of a specified translocation is denoted by T_{iSk}. This represents the component in the kth coordinate direction of the ith specified translocation. The general form of a component of working translocation is T_{iWk}. The notation is similar to that of T_{iSk} except for the subscript w (for working) that appear instead of s (for specified). The relationships between the two sets of components for the example of Fig. 6(a) may be written as follows:

$$T_{1W1} + T_{2W1}(10^{-Cd_{12}}) = T_{1S1}$$
$$T_{1W2} + T_{2W2}(10^{-Cd_{12}}) = T_{1S2}$$
$$T_{2W1} + T_{1W1}(10^{-Cd_{12}}) = T_{2S1}$$
$$T_{2W2} + T_{1W2}(10^{-Cd_{12}}) = T_{2S2}$$

where d_{12} is the 'relative distance' between the positions of the first and second specified translocations. The first equation states that the combination of

- the first component of working translocation at node 5 and
- the translocation in the first direction of node 5 caused by the first component of the working translocation at node 6

must add up to the first component of the specified translocation at node 5. The remaining three equations have similar implications.

The above equations in matrix notation will assume the form:

$$\left[\begin{array}{cc:cc} 1 & 0 & E & 0 \\ 0 & 1 & 0 & E \\ \hdashline E & 0 & 1 & 0 \\ 0 & E & 0 & 1 \end{array}\right] \left[\begin{array}{c} T_{1W1} \\ T_{1W2} \\ \hdashline T_{2W1} \\ T_{2W2} \end{array}\right] = \left[\begin{array}{c} T_{1S1} \\ T_{1S2} \\ \hdashline T_{2S1} \\ T_{2S2} \end{array}\right]$$

where

$$E = 10^{-Cd_{12}}$$

and since

$$d_{12} = \frac{3\sqrt{2}}{6\sqrt{2}} = \frac{1}{2}$$

then

$$E = 10^{-C/2} = \frac{1}{10^{C/2}}$$

The above matrix equation may be represented as

$$\mathrm{A\ T_w = Ts}$$

where A is referred to as the 'interaction matrix', T_w is referred to as the 'working translocation vector' and T_s is referred to as the 'specified translocation vector'. With the actual values of the components of the specified translocations inserted in T_s, the above matrix equation becomes:

$$\left[\begin{array}{cc:cc} 1 & 0 & E & 0 \\ 0 & 1 & 0 & E \\ \hdashline E & 0 & 1 & 0 \\ 0 & E & 0 & 1 \end{array}\right] \left[\begin{array}{c} T_{1W1} \\ T_{1W2} \\ \hdashline T_{2W1} \\ T_{2W2} \end{array}\right] = \left[\begin{array}{c} 2 \\ 1 \\ \hdashline 1 \\ -1 \end{array}\right]$$

The solution is found to be:

$$T_{1W1} = \frac{2-E}{1-E^2}, \quad T_{1W2} = \frac{1+E}{1-E^2}, \quad T_{2W1} = \frac{1-2E}{1-E^2} \quad \text{and} \quad T_{2W2} = -\frac{1+E}{1-E^2}$$

With the value of the control parameter C chosen as 1,

$$E = \frac{1}{\sqrt{10}}$$

and the working translocation components are found to be:

$T_{1W1} = 1.870858$, $T_{1W2} = 1.462475$, $T_{2W1} = 0.408383$ and $T_{2W2} = -1.462475$

It should be noted that the above system of simultaneous equations has a solution if and only if $E \lessgtr 1$, that is, if and only if $C \lessgtr 0$.

In general, an interaction matrix is a square symmetric matrix with a simple constitution. To illustrate the characteristic form of an interaction matrix, the equation

$$A\,T_w = T_s$$

for a three-directional configuration with specified translocations at four points is shown below:

$$
\left[\begin{array}{ccc:ccc:ccc:ccc}
1 & 0 & 0 & E_{12} & 0 & 0 & E_{13} & 0 & 0 & E_{14} & 0 & 0 \\
0 & 1 & 0 & 0 & E_{12} & 0 & 0 & E_{13} & 0 & 0 & E_{14} & 0 \\
0 & 0 & 1 & 0 & 0 & E_{12} & 0 & 0 & E_{13} & 0 & 0 & E_{14} \\
\hdashline
E_{12} & 0 & 0 & 1 & 0 & 0 & E_{23} & 0 & 0 & E_{24} & 0 & 0 \\
0 & E_{12} & 0 & 0 & 1 & 0 & 0 & E_{23} & 0 & 0 & E_{24} & 0 \\
0 & 0 & E_{12} & 0 & 0 & 1 & 0 & 0 & E_{23} & 0 & 0 & E_{24} \\
\hdashline
E_{13} & 0 & 0 & E_{23} & 0 & 0 & 1 & 0 & 0 & E_{34} & 0 & 0 \\
0 & E_{13} & 0 & 0 & E_{23} & 0 & 0 & 1 & 0 & 0 & E_{34} & 0 \\
0 & 0 & E_{13} & 0 & 0 & E_{23} & 0 & 0 & 1 & 0 & 0 & E_{34} \\
\hdashline
E_{14} & 0 & 0 & E_{24} & 0 & 0 & E_{34} & 0 & 0 & 1 & 0 & 0 \\
0 & E_{14} & 0 & 0 & E_{24} & 0 & 0 & E_{34} & 0 & 0 & 1 & 0 \\
0 & 0 & E_{14} & 0 & 0 & E_{24} & 0 & 0 & E_{34} & 0 & 0 & 1
\end{array}\right]
\left[\begin{array}{c}
T_{1W1} \\ T_{1W2} \\ T_{1W3} \\ T_{2W1} \\ T_{2W2} \\ T_{2W3} \\ T_{3W1} \\ T_{3W2} \\ T_{3W3} \\ T_{4W1} \\ T_{4W2} \\ T_{4W3}
\end{array}\right]
=
\left[\begin{array}{c}
T_{1S1} \\ T_{1S2} \\ T_{1S3} \\ \hdashline T_{2S1} \\ T_{2S2} \\ T_{2S3} \\ \hdashline T_{3S1} \\ T_{3S2} \\ T_{3S3} \\ \hdashline T_{4S1} \\ T_{4S2} \\ T_{4S3}
\end{array}\right]
$$

where

- $E_{12} = 10^{-Cd_{12}}$, $E_{13} = 10^{-Cd_{13}}$, ... etc,
- $d_{12} = \dfrac{D_{12}}{D}$, $d_{13} = \dfrac{D_{13}}{D}$, ... etc,
- D_{12} is the distance between the point that has the first specified translocation and the point that has the second specified translocation. D_{13} is the distance between the point that has the first specified translocation and the point that has the third specified translocation, ...etc and
- D is the length of the diagonal of the box frame of the configuration.

The number of coordinate directions in the above matrix is taken to be three. However, the concept of novation is applicable to configurations with any number of coordinate directions.

When there are two or more specified translocations then the control parameter C is required to be nonzero. This is due to the fact that when $C = 0$ then the interaction matrix A will be singular and the system of simultaneous equations $AT_W = T_S$ will not have a unique solution (With $C = 0$, all the nonzero off-diagonal elements of A will be equal to 1 and the matrix will have a number of identical rows and columns and a vanishing determinant). However, when there is only one specified translocation then the control parameter C may have a zero value without any problem. In this case every node of the configuration will undergo a translocation identical to the specified translocation.

When the system of simultaneous equations $AT_W = T_S$ has a unique solution then an efficient way of obtaining this solution will be to solve the following equivalent system of simultaneous equations:

$$\begin{bmatrix} 1 & E_{12} & E_{13} & E_{14} \\ E_{12} & 1 & E_{23} & E_{24} \\ E_{13} & E_{23} & 1 & E_{34} \\ E_{14} & E_{24} & E_{34} & 1 \end{bmatrix} \begin{bmatrix} T_{1W1} & T_{1W2} & T_{1W3} \\ T_{2W1} & T_{2W2} & T_{2W3} \\ T_{3W1} & T_{3W2} & T_{3W3} \\ T_{4W1} & T_{4W2} & T_{4W3} \end{bmatrix} = \begin{bmatrix} T_{1S1} & T_{1S2} & T_{1S3} \\ T_{2S1} & T_{2S2} & T_{2S3} \\ T_{3S1} & T_{3S2} & T_{3S3} \\ T_{4S1} & T_{4S2} & T_{4S3} \end{bmatrix}$$

Returning to the example of Fig. 6, once the working translocations are in hand, a generalised form of the procedure used for the example of Fig. 5(a) can be employed to deal with multiple specified translocations. This generalised procedure is given by the formula:

$$T_k = \sum_{i=1}^{n} T_{iWk} (10^{-Cd_{ij}})$$

where

T_k is the kth component of translocation at a typical node j,
n is the number of points that have specified translocations,
T_{iWk} is the kth component of the ith working translocation,
dij is the relative distance of node j from the position of the ith specified translocation and is given by the ratio Dij/D, where,
Dij is the distance of node j from the position of the ith specified translocation,
D is the length of the diagonal of the box frame of the configuration and
C is the control parameter.

The application of this procedure to the configuration of Fig. 6(a) will give rise to the configuration shown by dotted lines, with the numerical results given in the table of Fig. 6(b).

7. Further Examples

Some examples of novational transformations are given in Fig. 7. The configuration shown in Fig. 7(a) is a planar configuration consisting of I8xI2 square elements. The configuration shown in Fig. 7(b) is obtained by a sharp novation of the configuration of Fig. 7(a) with the specified translocations being at the four corner nodes. These translocations are along the diagonals of the configuration. It is seen that the novation has only affected the four corner nodes that have specified translocations and all the other nodes are left in their original positions.

The configurations shown in Figs. 7(c), 7(d) and 7(e) are obtained using ED novations with the same specified translocations as in Fig. 7(b). The difference between the novations that have resulted in the configurations of Figs. 7(c), 7(d) and 7(e) is in the value of the control parameter *C*. The values of *C* for Figs. 7(c), 7(d) and 7(e) are 2, 1 and 6, respectively. The value of *C* controls the 'spread' of the influence of the specified translocations. The higher the value of *C*, the greater the 'decay' in the spread of the influence of the specified translocations will be. That is, the higher the value of *C*, the more localised the influence of the specified translocations will be. In other words, the lower the value of *C*, the larger the conformity translocations will be. That is, the lower the value of *C*, the more far reaching the influence of the specified translocations will be.

The configurations of Figs. 7(f), 7(g) and 7(h) are obtained using ED novations with the control parameter *C* being equal to 1. The specified translocations for these configurations are at the mid-points of the edges. In Fig. 7(f), the translocations are outward in both UI and U2 directions. In Fig. 7(g), the translocations are outward in the U1 direction and inward in the U2 direction. In Fig. 7(h), the translocations are inward in both U1 and U2 directions.

Some examples of novational transformations in a three-directional space are shown in Fig. 8. The configuration in Fig. 8(a) consists of an array of I6x24 square elements lying in the UI-U2 plane. The configurations in Figs. 8(b) to 8(h) are obtained using ED novations with the control parameter C being equal to 1.

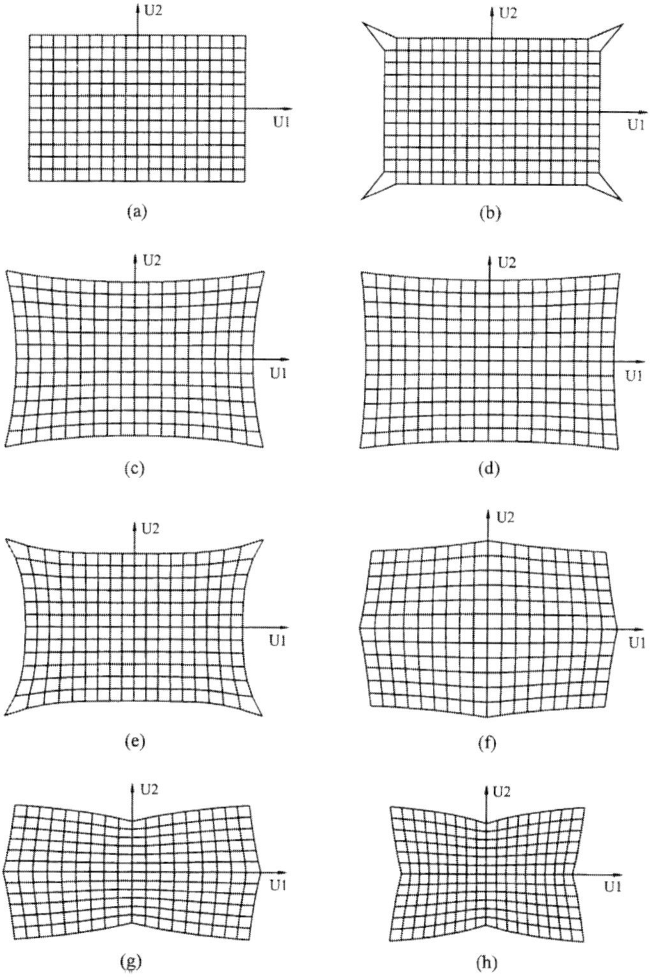

Fig. 7. Examples of planar ED novations.

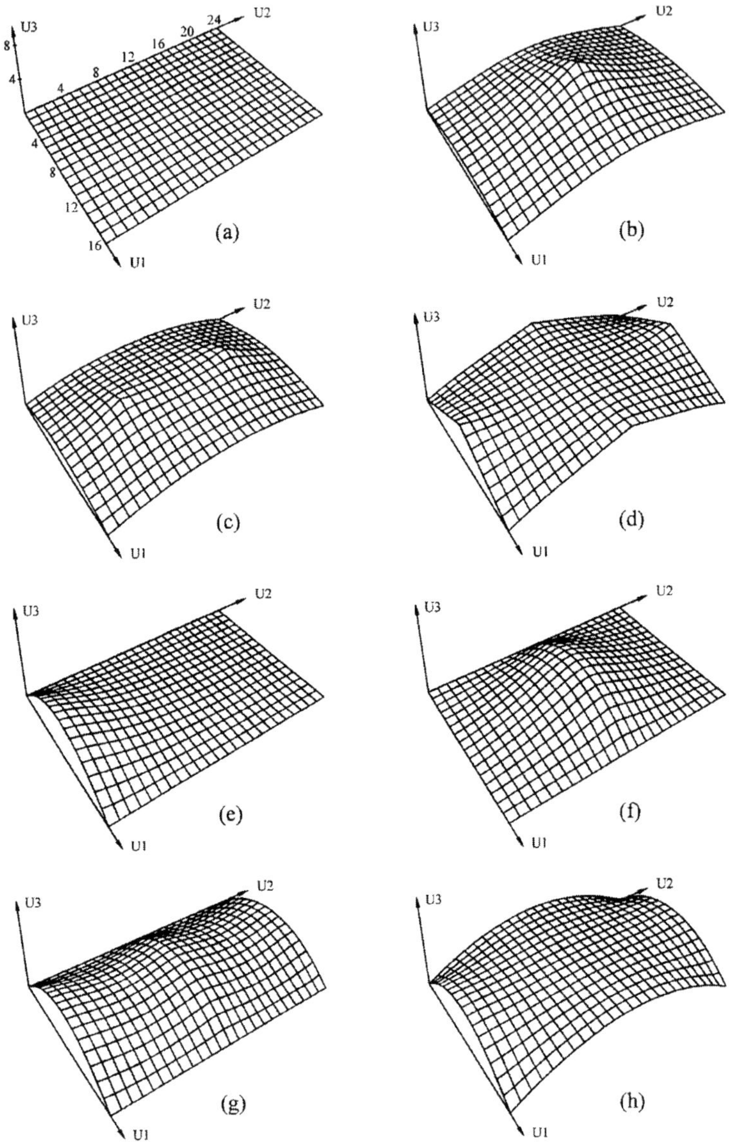

Fig. 8. Examples of spatial ED novations.

The configurations of Figs. 8(b), 8(c) and 8(d) are obtained by holding the four corner nodes in their original positions and raising one or more of the other nodes in the U3 direction. In Fig. 8(b), the central node is raised by 6 units. In Fig. 8(c), two internal nodes are raised by 4 units. In Fig. 8(d), the mid-points of the edges are raised by 4 units.

Fig. 8(e) is obtained by creating a parabolic arch along one edge while holding all the nodes along the other three edges at their original positions. The specification of the translocations that produces the parabolic arch is achieved using the equation

$$U3 = \frac{4hU1}{L} - \frac{4hU1^2}{L^2}$$

where L is the span of the arch (16 in the present example) and h is the height of the arch at the middle (4 in the present example). The actual specification of the translocations for this example is given in section 9.

The configuration of Fig. 8(f) is obtained in a similar manner except for the position of the arch which is along a central line. The same approach is used for the creation of Fig. 8(g) that involves three arches. Finally, the configuration of Fig. 8(h) is obtained by creating arches along all four edges.

8. Indirect ED Novations

In the case of a sharp novation, the specified translocations are required to be at nodal positions of the configuration. However, as far as an ED novation is concerned, the specified translocations need not necessarily be at nodal positions. An ED novation in which all the specified translocations are at nodal positions is referred to as a 'nodal ED novation' or a 'direct ED novation'. In contrast, an ED novation that involves one or more specified translocations at non-nodal positions is referred to as a 'non-nodal ED novation' or an 'indirect ED novation'. The process of ED novation as described in the paper will be identical for both direct and indirect ED novations except for a minor difference that for a direct ED novation the relative distance d_{ij} is always in the range 0 to 1 but in the case of an indirect ED novation the value of d_{ij} may be greater than 1.

Operation Mode: This is an integer expression whose value is either 1 or 2, where:
m=1 indicates that the operation to be performed is a sharp novation and
m=2 indicates that the operation to be performed is an ED novation.

Control Parameter: This is a numeric expression whose value will be used as the control parameter for an ED novation. The presence of C is optional. If m=2 and C is absent then the control parameter is assumed to be equal to 1. On the other hand, if m=1 and C is present then it is ignored.

G is a formex variable that represents the novation of the configuration which is represented by E. If E={} then G={}.

Argument: E is a formex expression whose value represents the configuration to be novated. E may represent the empty formex but if m=2 then E should not represent a configuration whose box frame is a single point.

$$G = NOV(m, C, L1, L2) \mid E$$

Abbreviation for novation

First List: This is a formex expression whose value is an ingot of the form:
$\{[U_{11}, U_{12}, \ldots, U_{1r}], [U_{21}, U_{22}, \ldots, U_{2r}], \ldots, [U_{n1}, U_{n2}, \ldots, U_{nr}]\}$,
where each signet represents the coordinates of a point that has a specified translocation and where r is the number of coordinate directions and n is the number of points that have specified translocations. The following conditions must be satisfied:

(1) Every signet of L1 must be unique. That is, a point should not be represented more than once in L1.
(2) The number of coordinate directions in L1 (that is, r) must be the same as that in the argument E.
(3) In the case of a sharp novation, every signet of L1 must be contained in the argument E.

Second List: This is a formex expression whose value is an ingot of the form: $\{[V_{11}, V_{12}, \ldots, V_{1r}], [V_{21}, V_{22}, \ldots, V_{2r}], \ldots, [V_{n1}, V_{n2}, \ldots, V_{nr}]\}$, where the signets represent the translocated coordinates of the points represented in L1, in the same order as in L1. The components of the specified translocation at a point are obtained as the differences between the translocated coordinates of the point, as given in L2, and the corresponding initial coordinates of the point, as given in L1. For instance, the components of the second specified translocation are obtained as follows:
$T_{2S1}=V_{21}-U_{21}, T_{2S2}=V_{22}-U_{22}, \ldots, T_{2Sr}=V_{2r}-U_{2r}$

Fig. 10. Particulars of novation function.

Details regarding the formulations, notation and terminology used in the above example (as well as in Fig. 10) are given in Reference 1.

The novation function has been in use in formex configuration processing since the mid-eighties.[2] However, the definition of this function has been in a form equivalent to what is called 'sharp novation' in the present work. The extended definition of the novation function, as described in this paper, has been recently implemented in Formian and is available for use.

Acknowledgements

The help of Taiyo Kogyo Corporation and Tomoe Corporation in supporting the development of Formian is gratefully acknowledged.

References

1. Nooshin, H. and Disney, P. L., Formian 2, Multi-science Publishing, London, 1997.
2. Nooshin, H., Formex Configuration Processing in Structural Engineering, Elsevier Applied Science Publishers, London, 1984.

CHAPTER 6

SOME STRUCTURAL-MORPHOLOGICAL ASPECTS OF DEPLOYABLE STRUCTURES FOR SPACE ENCLOSURES

Ariel Hanaor

Sr. Research Associate, National Building Research Institute, The Technion, Israel Institute of Technology, Technion City, Haifa 32000, Israel

The paper is a review of deployable structural systems that have been proposed recently for the purpose of space enclosure. The structural systems are characterized and classified by their structural-morphological properties and by the kinematics of deployment. Some retractable and dismountable configurations are also reviewed. The systems are evaluated in terms of their structural efficiency, technical complexity and deployment/stowage efficiencies. The paper includes an extended list of references.

1. General Principles

1.1. *Definitions and Scope*

Deployable structures are generally used in two types of applications: a) as temporary structures; b) in inaccessible or remote places, such as outer-space. The first application implies a reversible process of deployment and undeployment, while the second may not. Two extreme types of deployable structures can be distinguished: Fully deployable structures are fully assembled in the stowed state, and the deployment process involves no component assembly. In fully dismountable structures (assemblable according to Ref. 9), on the other hand, the structure is stowed as separate components (at member level). It is assembled on site from these components and can be disassembled back into the stowed state.

The paper deals primarily with structures that are fully deployable or composed of large deployable sub-assemblages, and with the specific application of space enclosure. However, some reference is made to some types of dismountable structures and also to certain types of retractable roofs. Although outer-space applications are excluded from the scope, some technologies that have been proposed for outer-space applications are adaptable to terrestrial space enclosures and are included in this review and in the reference list.

1.2. *Classification*

The purpose of a classification system is to highlight in a hierarchical fashion the principles governing the set of objects under discussion. The present paper is concerned with the structural-morphological properties as the primary interest. The overall classification system is presented in the chart in Fig. 1, in the form of a two-way table. The columns of the table represent the morphological aspects and the rows the kinematic properties, which are of primary significance in the context of deployable structures.

Two subcategories are considered for each of the main classification categories. The two major morphological features are lattice or skeletal structures, and continuous or stressed-skin structures. It should be noted that in the context of space enclosures, all structures have a functional covering surface. The difference between the two classes of structures mentioned above is that in skeletal structures, the primary load-bearing structure consists of discrete members, whereas in continuous structures the surface covering itself performs the major load-bearing function. A third class, namely hybrid structures, combines skeletal and stressed-skin components with approximately equal roles in the load-bearing hierarchy, but in the present classification each of the components is dealt with in its respective class. The two major kinematic subcategories are systems comprised of rigid links, such as bars or plates, and systems containing deformable or soft components, which lack flexural stiffness, such as cables or fabric. In general, the deployment of structures composed of rigid links can be more accurately controlled than that of deformable structures, but usually at a cost of increased mechanical complexity.

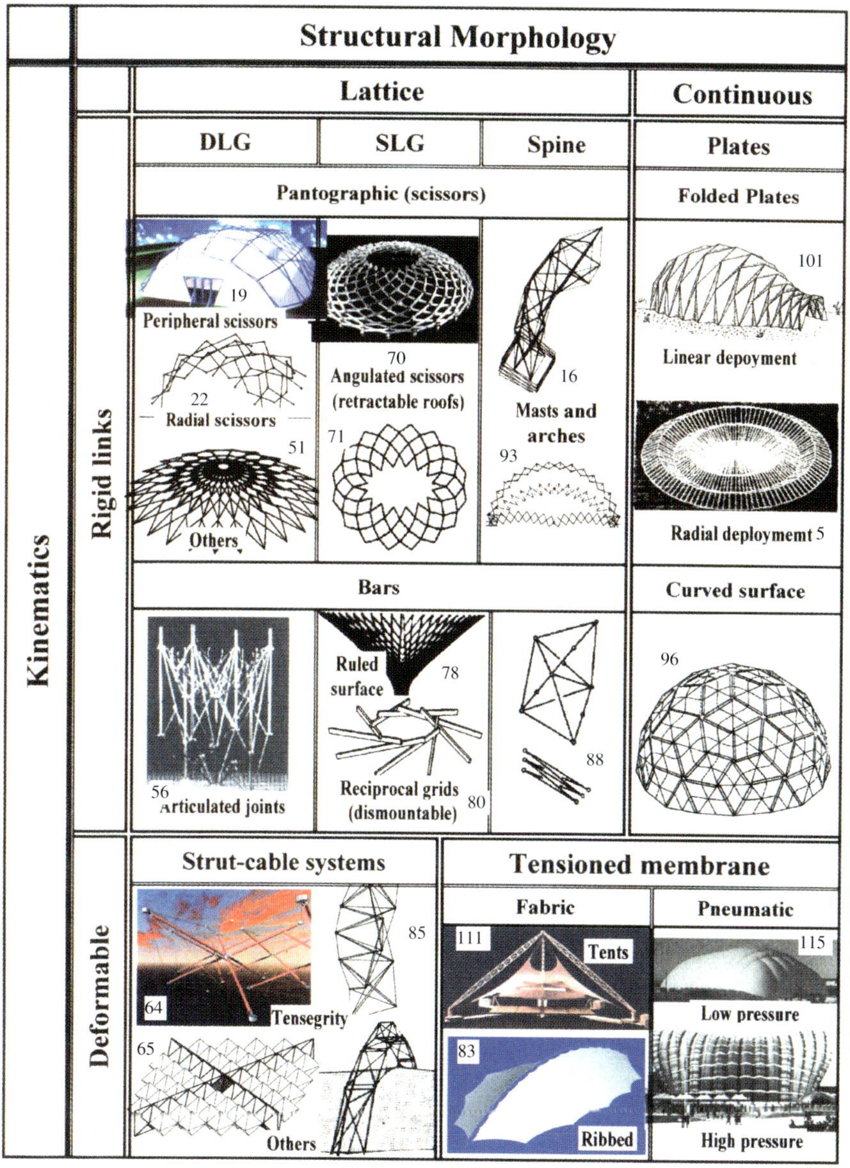

Fig. 1. Deployable structures classification chart. Numerals indicate illustration references.

Each of the four major classes generated by this two-way division is further subdivided according to more specific morphological features, as indicated in Fig. 1 and discussed in more detail below. The extended reference list at the end of the paper is arranged in accordance with the classification scheme. References 1-13 deal with topics common to several differing systems.

1.3. *Kinematics*

Kent[5] defines a kinematic structure as one having a single kinematic degree of freedom (KDOF), namely, the positioning of one node in the structure relative to the others, determines uniquely the geometry of the structure. Such systems represent the ultimate deployment control, since only one point needs to be controlled to determine the geometry at any instant during the deployment. While such feature is of great advantage in outer-space applications, it is of lesser significance in terrestrial applications, where several degrees of freedom are normally constrained (e.g. along the boundaries) and where manual intervention is feasible. Obviously, such kinematic control is possible only in structures consisting of rigid links, since deformable components have infinite degrees of freedom. The majority of deployable space enclosure structures are based on hinged mechanisms. The "minimal" hinged mechanism possessing a single kinematic degree of freedom (SKDOF) is one composed of four links (members) and four hinges – Fig. 2a. This KDOF can be propagated in the plane or in space by extending the links and connecting units to generate periodic structures (Fig. 2c presents a simple planar example). In space, more complex basic mechanisms are possible,[5] but in practice most deployable structures, particularly in terrestrial applications, consist of combinations of planar mechanisms. The constraint of SKDOF and the limited range of basic mechanisms notwithstanding, the range of kinematic configurations that can be constructed is vast, although only part of the configurations presented in Ref. 5 are applicable for deployable structures.

Fig. 2. Planar SKDOF mechanisms: a) Basic minimal mechanism with rotational releases; b) Basic mechanism with slide releases; c) Propagation of basic rotational mechanism – pantographs.

1.4. *Design and Evaluation Criteria*

General design criteria for deployable structures are derivatives of the function of these structures as temporary, possibly multi-purpose and of repeated use. Design criteria relate generally to three phases in the structure's life cycle. These phases are: design; storage and haulage; site operations (deployment/dismounting). Structural-morphological (including kinematic) considerations affect and are associated with all design criteria to a larger or lesser extent. Howeveer, The present paper is concerned primarily with structural-morphological aspects of deployable structures, rather than with the deployment process and its efficiency, and the discussion that follows is focused on these aspects. A primary structural criterion is structural efficiency, defined as the ratio of load bearing capacity to weight (for a given material). The primary parameter affecting *structural efficiency* is *structural depth*, which controls the magnitude of internal forces and stresses. The review is arranged in the sequence of the classification chart (Fig. 1), where each system is described in terms of its main morphological, kinematic and structural characteristics.

2. Double-Layer Grids

2.1. *Pantographic Grids*

Morphology and *kinematics*: Double-layer pantographic grids are mostly based on prismatic units composed of scissors. These units are connected at the apices or along the sides of their base polygons to generate tessellations, with vertices lying in two parallel surfaces connected by scissor units. There are no bars actually lying in the surfaces to form chords. Two basic types of scissor units can be identified – Fig. 3: The more common type of grid is composed of units of peripheral scissors, where each side face of the basic prism is formed by a pair of scissors (Fig. 3a). The less common type is composed of radial scissors, where scissor-arms intersect with a common composite hinge at the centroid of the prism (Fig. 3b). It is possible to generate basic units that combine peripheral and radial units,[50] but these are too complex mechanically, to be of practical use. Some composite basic units have been proposed in the literature,[30,31,42,43] which combine several peripheral triangular prisms in a "clicking" mechanism, providing improved stiffness and a locking device. Such a unit, composed of four right-angle triangles, is shown in Fig. 3c.

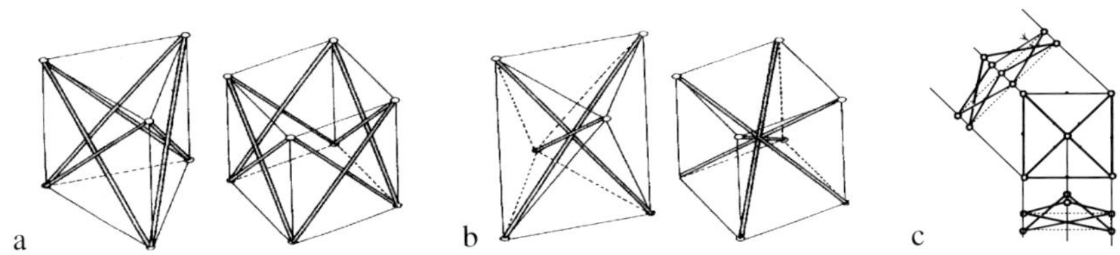

Fig. 3. Basic scissors units: a) Peripheral scissors – triangular and rectangular prisms[22]; b) Radial scissors – triangular antiprism, rectangular prism[22]; c) A "clicking" unit.[43]

Configurations based on triangular and rectangular or rhomboid units, are the most common that have been proposed to date for actual implementation, but a large variety of geometries can be found in the literature. In the chart in Fig. 1 are shown a structure based on rectangular peripheral units[19] and one based on triangular radial units connected at their vertices (with "holes" between units).[22] The latter configuration, favored by Piñero[36] has a low bar density but also low stiffness. Some configurations have been proposed in the literature, which are not based on propagation of basic units. These are mostly grids with radial deployment.[5] In the chart of Fig. 1 is shown a domical configuration based on radial and ring scissors.[51]

The kinematic motion of the scissors is propagated along the two-way or three-way pantograph as a single degree of freedom. However, structures forming non-triangular grids have additional in-plane degrees of freedom, implying that the generated grid is not geometrically rigid and requires stiffening of the surface, particularly when the surface is curved. A doubly curved triangular grid, on the other hand, is geometrically rigid but kinematically incompatible, except at the deployed and the folded states. At intermediate states, there is some resistance to motion, depending on the stiffness of the components and joints. A "clicking" effect is thus generated in triangular grids.

Structural efficiency: The pantograph on its own lacks structural depth, beyond the depth of the individual bars, and therefore would have an extremely low structural efficiency as a planar surface. The double-layer aspect is a geometrical, not a structural feature, unless chords are added in the form of additional bars or tendons. This is the rcason that curved surfaces are favored, providing a structural depth and enabling direct (axial) action as the primary mode. Flexure in the bars as a secondary action nevertheless remains a major feature that detracts from structural efficiency. The need for articulation at joints reduces stiffness and introduces a source of imperfections. The surface membrane, commonly employed in these structures, can be used as a bracing element, as well as to control member buckling. Unfortunately, very little attention has been given to this topic. Such treatment as exists has been limited primarily to geometric and kinematic compatibility aspects.[23]

2.2. *Other DLG Bar Systems*

It appears that the pantograph principle, or variations on it, is the simplest technique for producing deployable bar structures. Few alternative concepts for double-layer grids have been proposed that consist entirely of bars, and the majority of those were developed for outer-space applications. The main motivation for these concepts is the high degree of reliability in deployment required in outer space. These concepts often involve complex articulated joints and mechanisms, including sliding joints (see for instance illustration in Fig. 1[56]). Adaptation of these concepts to terrestrial applications is feasible, but even when simplifications are introduced, such concepts are expected to be costly and limited to highly specialized applications.

2.3. *Strut-Cable Systems*

Several deployable concepts have been proposed in the literature, which combine rigid bars with cables in a variety of configurations. The principle has high application potential in that it enables to optimise between the sometimes conflicting requirements of deployment reliability, technical simplicity (limitation of articulated mechanical joints) and structural efficiency (weight reduction).

Morphology and kinematics: This class of structures combines together a number of morphologically and topologically distinct systems. The predominant concept in this group is the so-called "*tensegrity*" structures. The term is ill defined and widely differing concepts have been proposed in the literature under this catchy label. In its "pure" form, a tensegrity structure consists of a network of bars and cables, in which any bar is connected only to cables and to no other bar. Such restrictive definition is important in the context of deployable structures because it implies the complete absence of articulated joints. The deployability is provided by the deformable cables. The deployment process of such structures requires the change in length of members. Accordingly, two deployment techniques have been proposed. One technique involves the change of bar length by means of energy supply (hydraulic or mechanical) to telescoping bars. The other technique involves the pulling of cables over a system of pulleys attached to the bars. Each method has its advantages and drawbacks regarding mechanical complexity, deployment reliability and structural efficiency. A generalized tensegrity concept can be defined as having a disconnected (in the graph-theoretical sense) system of bars, namely, consisting of disjoint groups of connected bars. In this case some articulation is required. The illustration of the tensegrity concepts in Fig. 1[64,85] is an example.

Other concepts involving combinations of bars and cables have been proposed. The second illustration of a DLG system in Fig. 1 involves pantographic spines that in their deployment prestress a network of octahedra consisting of bars and cables.[65] The system has a single KDOF and it was proposed for outer-space applications, but it is readily implementable in terrestrial applications. The configuration shown lacks substantial chords. To achieve structural depth, curved surfaces need to be generated, or chords added (in the form of either bars or prestressed tendons).

Structural efficiency: Structural efficiency of pure tensegrity structures is governed by the long unbraced bars. Structural efficiency of configurations with bars joined between the two layers, as in the configuration in Fig. 1,[64] is considerably improved. Employment of the cover membrane to brace struts in configurations which facilitate this can further improve efficiency.

3. Single-Layer Retractable and Dismountable Grids

3.1. *Angulated Pantographic Grids*

Two types of pantographic angulated scissors **retractable roofs** are illustrated in Fig. 1. The better known Hoberman concept[70] involves individual "two-armed" scissors, whereas the second type[71] involves continuous rigid angulated arms curving from the centre of the circular or oval roof (in the closed position) towards the perimeter in two crossing families. This latter configuration can be surfaced with rigid covering panels attached to one of the families of rigid arms (one panel per arm). During closing and opening each of these arms rotates around a fixed point.

3.2. *Reciprocal Grids and Ruled Surfaces*

Reciprocal grids are essentially domical surfaces consisting of mutually supported beams. As such they are subjected to considerable flexure and therefore possess low structural efficiency. Configurations consisting of non-triangular cells (usually rhomboid cells) are theoretically foldable to a bundle, but in practice the members are long and bulky and the bundles are not very compact. For this reason the concept is suitable for dismountable rather than deployable applications. The surface can be assembled simply and rapidly from the individual members using very simple joints (seated joints are sufficient). The surface is suitable for rigid covering, which can serve both for bracing and for improving structural efficiency, but the attachment of this surface detracts from the speed of erection of the structure.

A large variety of ruled hyperbolic surfaces and composites of such surfaces can be generated, consisting of two families of straight members.[5,78] Theoretically, these surfaces, which consist of rhomboid cells, fold to a bundle but in practice they are subject to the same limitation as reciprocal grids and are suitable for dismountable, rather than deployable structures.

4. Masts and Spines for Fabric and Hybrid Structures

4.1. *Rigid Bar Grids*

Pantographic grids: A pantographic spine is produced from any of the basic prismatic scissor units (Fig. 3) by joining them not side by side, but at the prism bases. Straight or arched configurations can be generated.[16] These configurations do not fold to a bar bundle but to polygonal stacks. Furthermore, the joints involved are mechanically more complex than the essentially planar joints of the DLG configurations. Planar pantographs can also be used, which fold into a bar bundle. Such configurations rely on the surface membrane to restrain lateral buckling.[91,93]

Other bar grids: Several concepts employing articulated joints have been proposed,[86,88] particularly for booms in outer space, but these often involve complex articulated joints and are thus very costly. However, it should be borne in mind that spine elements are few and far between, so the cost is spread over the large area covered. High stiffness and strength are often required from these structures due to the high loads they may be subjected to and such requirements may not be satisfied by pantographic or strut-cable systems.

4.2. *Strut-Cable Systems*

As observed for DLGs, pure tensegrity structures possess low structural efficiency and low stiffness, and therefore they are unsuitable for use as spines, which collect loads from a large area. Other types include generalized tensegrity grids, namely involving strut-strut contact[83,85] and configurations that combine rigid bar units with cables as means of deployment and prestress.[92] A super-light application[84] for fabric structures involves spines consisting of chords made of fiber composite rods and fabric web (rather than cables). These concepts are often semi-deployable or dismountable,[83,84,92] but since few elements are involved, the deployability of the system as a whole is not significantly affected.

5. Plate Structures

5.1. *Folded Plates*

Kinematics and Morphology: The basic "minimal" SKDOF mechanism of plate structures consists of four plates connected by hinges. Two configurations can be distinguished as shown in Fig. 4a, one consisting of parallel hinges and the other with hinges intersecting at a point. The parallel fold gives rise to planar folded surfaces whereas the intersecting fold enables curved configurations. Both configurations fold to (theoretically) flat planes in two ways, provided certain geometric relations between the links are observed.[5] These basic units can be combined to generate a very large array of folded surfaces.[5,94] These "four-fold" configurations (Fig. 4b) have a single KDOF. In practice,

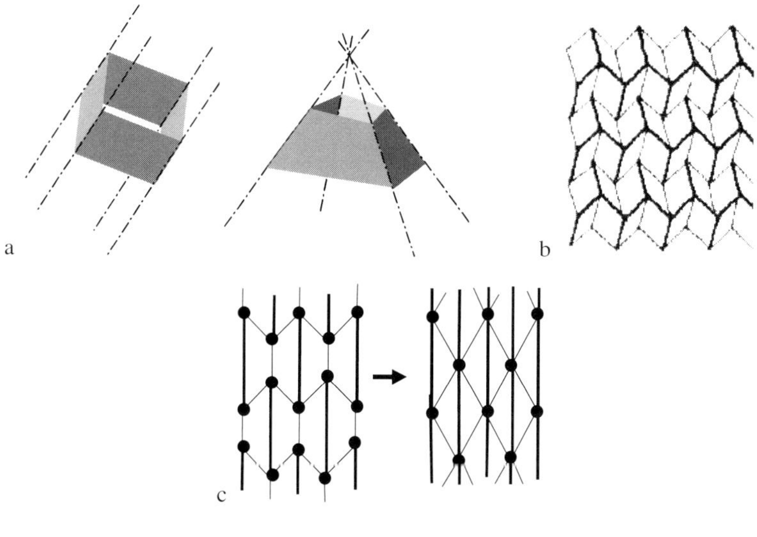

Fig. 4. a) Basic "minimal" SKDOF mechanisms for generating folded plate structures[5]; b) Four-fold folded surface[94]; c) Merging of four-fold to six-fold pattern.

folded surfaces are usually generated with six folds, rather than four, meeting at a point (see, for instance illustration of linear deployment in Fig. 1[101]). This configuration is, in fact, a merging of two four-fold vertices – Fig. 4c. This merging enables increasing the surface curvature but results in increasing also the number of KDOFs. This is not a major drawback, since definition of the boundaries uniquely defines the geometry and provides control of deployment.

The majority of folded plate structures proposed to date have a linear (or curvilinear) deployment and they fold into a compact slab of stacked plates. Some configurations have been proposed in the literature that fold to a compact cylinder and deploy radially out of the cylinder to form a folded "disc".[5,97] The illustration in Fig. 1[5] shows such a disc with circumferential folds in the deployed state (domical configurations are also possible). To fold it, a radial "cut" is made and the edges overlap (several turns, theoretically) until a compact spiral cylinder is produced. To realize such configurations with curved folds in practice, the material has to be very thin and flexible. A configuration with spiral folds that folds to a compact cylinder has been proposed[97] originally for deployment of membrane in outer space, but it can also be realised as a folded plate. This pattern does not involve overlap and is therefore more suitable for implementation as deployable structure.

Structural efficiency and weight: Folded plate structures inherently possess high structural efficiency. This inherent efficiency, however, does not automatically translate into light weight, since the plates themselves, which are subject to compression and flexure, require minimum dimensions. The resulting overall weight may be higher than in structures surfaced with membrane.

5.2. *Curved Surfaces*

Doubly (and singly) curved surfaces can be discretised into plate elements. In general, these surfaces cannot be folded and deployed as one unit. Unfolded planar patterns can be obtained by disconnecting certain joints between plates and these patterns can, in principle, be further folded into compact stacks. The concept is more suitable for dismountable or semi-dismountable implementation than to primarily deployable ones – See illustration in Fig. 1.[96]

6. Tensioned Membrane Systems

6.1. *Fabric and Hybrid Structures*

Morphology and kinematics: As a rule, a pure tension structure does not exist. To maintain equilibrium some compressive elements must be present. In fabric structures these elements are composed of rigid bars, singly or as skeletal elements (in pneumatic structures the compressive element is primarily air pressure). Two main types of fabric structures are distinguished: In *tents*, the compression elements, usually in the form of masts, are external to the fabric surface and act separately from it. In *ribbed structures*, the compression elements form part of the surface, and usually interact structurally with the membrane. In *hybrid structures* the compression element is a self-supporting structure separate from the membrane and supporting it. In this sense these are different from the two other types, where the compression elements are both supporting and are supported (or braced) by the membrane.

Since the membrane itself is inherently deployable, the kinematics and deployability are governed by the deployability of the compressive elements. This aspect is discussed under section 4 above – "Masts and spines for fabric and hybrid structures". The illustrations in Fig. 1 show a deployable structure that can be considered a tent or a hybrid structure,[111] and a ribbed fabric structure.[83] The tent is used as a cover for a mobile stage and each of the three external masts folds in half and occupies a semi-trailer in transportation. The low stowage efficiency is made up for by rapid deployment. The ribbed structure employs a dismountable strut-cable spine.

Structural efficiency: Structural efficiency is extremely high when adequate structural depth is provided by the surface. Overall weight is governed by the compression elements rather than by the surface covering. Ribbed structures, in which the membrane restrains lateral and overall buckling, are particularly efficient and light. In general, fabric

8. Kronenburg, R., Portable Architecture, 2nd ed., Butterworth-Heinemann, 1999.
9. Levy, M. and Luebkeman, C., "Typology of Mobile and Rapidly Assembled Structures", Shells and Spatial Structures: From Recent Past to the Next Millenium, Proc. IASS 40th Anniversary congress, Madrid, 1999, V II, D13.
10. Miura, K., "Concepts of Deployable Space Structures", Int. J. Space Structures, Special Issue on Deployable Space Structures, V8 N1&2, 1993, 3-16.
11. Pellegrino, S., "Deployable Structures", Current and Emerging Technologies of Shell and Spatial Structures, Proc. IASS Col., Able, J. Astudillo, N.K. and Srivastava, N.K. eds., Madrid, April 1997, 175-188.
12. Robbin, T., "Deployable Structures", Engineering a New Architecture, Ch. 4, Yale University Press, New Haven, 1996, 38-57.
13. Srivastava, N.K. and LeBlanc, M., "Retractable Roof Structures", Current and Emerging Technologies of Shell and Spatial Structures, Proc. IASS Coll., Madrid, April 1997, 91-104.

7.2. *Double-Layer Pantographic Grids*

14. Al Khayer, M. and Lalvani, H., "Scissors-Action Deployables Based on Space-Filling of Polygonal Hyperboloids", Deployable Structures: Theory and Application, Book of Abstracts, IUTAM/IASS Symp., Cambridge, U.K., September, 1998.
15. Clarke, R.C., "The Kinematics of a Novel Deployable Structure System", Proc. 3rd Int. Conf. Space StructuresI, University of Surrey, Sept. 1984, Elsevier Science Publishers, London, 820-822.
16. Escrig, F., "Expandable Space Frame Structures", Proc. 3rd Int. Conf. Space Structures, University of Surrey, Sept. 1984, Elsevier Science Publishers, London, 845-850.
17. Escrig, F., "Expandable Space Structures", Int J. Space Structures, 1, 2, 1985, 79-91.
18. Escrig, F., "A New Geometry for Cylindrical Deployable X-Frames", Structural Morphology, Towards the New Millennium, Proc. Int. Coll. IASS, Univ. of Nottingham, Aug. 1997, 332-339.

19. Escrig, F., personal communication.
20. Escrig, F. and Valcarcel, J.P., "Analysis of Expandable Space Bar Structures", Shells, Membranes and Space Frames, Proc. IASS Symp., Osaka, Sept. 1986, V3, 269-276.
21. Escrig, F. and Valcarcel, J.P., "Curved Expandable Space Grids", Proc. Int. Conf. Design and Construction of Non-conventional Structures, Topping, B.H.V., ed., Civil Comp Press, Edinburgh, 1987, 157-166.
22. Escrig, F. and Valcarcel, J.P., "Geometry of Expandable Space Structures", Int. J. Space Structures, 8, 1-2, 1993, 71-84.
23. Escrig, F., Valcarcel, J.P. and Sanches, J., "Roofing Geometry of Deployable X-Frames", Int. J. Space Structures, 13, 1, 1998, 1-12.
24. Escrig, F., Sánchez, J. and Valcárcel J.P., "Foldable Grids of Coupled Arches", Shells and Spatial Structures: From Recent Past to the Next Millenium, Proc. IASS 40th Anniversary congress, Madrid, 1999, V II, D41.
25. Escrig, F., Valcarcel, J.P. and Sanches, J., "Deployable Cover on a Swimming Pool in Seville", J. IASS, 37, 1, 1996, 39-70.
26. Gantes, C., "Geometric Constraints in Assembling Polygonal Deployable Units to Form Multi-unit Structural Systems", Space Structures 4, Proc. 4th Int. Conf. on Space Structures, University of Surrey, 1993, Thomas Telford, London, V 1, 793-804.
27. Gantes, C., "Analytical Prediction of the Snap-through Characteristics of Deployable Structures", Mobile and Rapidly Assembled Structures II, Proc. Intnl Conf. MARAS 96, Computational Mechanics Publications, Southampton, 1996, 83-92.
28. Gantes, C., Deployable Structures, Analysis and Design, WIT Press, Southampton, UK, 1999.
29. Gantes, C., Connor, J.J. and Logcher, R.D., "A Systematic Design Methodology for Deployable Structures", Int. J. Space Structures, 9, 2, 1994, 67-86.
30. Gantes, C.J., Logcher, R.D., Connor, J.J. and Rosenfeld, Y., "Deployability conditions for Curved and Flat, Polygonal and Trapezoidal Deployable Structures", Int. J. Space Structures, V8, N1,2, 1993, 97-106.

31. Gantes, C.J., Logcher, R.D., Connor, J.J. and Rosenfeld, Y., "Geometric Design of Deployable Structures with Discrete Joint size", Int. J. Space Structures, Special Issue on Deployable Space Structures, 8, 1-2, 1993, 107-117.
32. Hoberman, C., "Art and Science of Folding Structures", Sites, V24, 1992, 34-53.
33. Kwan, A.S.K., "A Parabolic Pantographic Deployable Antenna (PDA)", Int. J. Space Structures, 10, 4, 1995, 195-203.
34. Langbecker, T., "Kinematic Analysis of Deployable Scissor Structures", Int. J. Space Structures, 14, 1, 1999, 1-15.
35. Pellegrino, S. and You, Z., "Foldable Ring Structures", Space Structures 4, Proc. 4th Int. Conf. on Space Structures, University of Surrey, September, 1993, Thomas Telford, London, V1, 783-792.
36. Piñero, E.P., "Teatro Ambulante", Arquitectura, N. 30, Junio, 1961, 27-33.
37. Piñero, E.P., Three Dimensional Reticular Structure, U.S. Patent 3.185.164, 1965.
38. Piñero, E.P., "Estructuras Reticulantes", Arquitectura, 112, April 1968, 1-9.
39. Piñero, E.P., "Estructures Reticulée", L'Architecture d'aujourd'hui, 141, Diciembre 1968, 76-81.
40. Piñero, E.P., "Teatros Desmontables", Informes de la Construcción, 231, 1971, 34-43.
41. Puertas del Rio, L., "Space Frames for Deployable Domes", Bull. IASS, 32, 2, 1991, 107-113.
42. Rosenfeld, Y. and Logcher, R.D., "New Concepts for Deployable Collapsable Structures", Int. J. Space Structures, 3, 1, 1988, 20-32.
43. Rosenfeld, Y., Ben-Ami, Y. and Logcher, R.D., "A Prototype 'Clicking' Scissor-Link Deployable Structure", Int. J. Space Structures, 8, 1-2, 1993, 85-95.
44. Sánchez-Cuenca, L., "Geometric Models for Expandable Structures", Mobile and Rapidly Assembled Structures II, Proc. Intnl Conf. MARAS 96, Computational Mechanics Publications, Southampton, 1996, 93-102.

45. Servadio, I., "Deployable Regular and Semi-regular Polyhedral Structures using Hoberman's Technique", IASS Bulletin, 36, 2, 1995, 91-97.
46. Shan, W., "Configuration Studies of Foldable Structures", Space Structures 4, Proc. 4th Int. Conf. on Space Structures, University of Surrey, September, 1993, Thomas Telford, London, V1, 824-832.
47. Valcárcel, J.B., Escrig, F. and Martin, E., "Expandable Domes with Incorporated Roofing Elements", Space Structures 4, Proc. 4th Int. Conf. on Space Structures, University of Surrey, September, 1993, Thomas Telford, London, V1, 805-812.
48. Valcárcel, J.P. and Escrig, F., "Recent Advances in the Analysis of Expandable Structures", Mobile and Rapidly Assembled Structures II, Proc. Intnl Conf. MARAS 96, Computational Mechanics Publications, Southampton, 1996, 46-54.
49. Valcárcel, J.P., Escrig, F., Cominguez, E. and Lamas, V., "Expandable Structures with Selffolding Stiffeners", Shells and Spatial Structures: From Recent Past to the Next Millenium, Proc. IASS 40th Anniversary congress, Madrid, 1999, V II, D21.
50. Wu, H.K. and Lalvani, H., "A Class of Periodic Deployable Space Frames using Regular Prismatic Nodes", Structural Morphology, Towards the New Millennium, Proc. IASS coll., Univ. of Nottingham, 1997, 76-83, Bull. IASS, 39, 1, 1998, 37-45.
51. You, Z., "A Pantographic Deployable Conic Structure" Int. J. Space Structures, 11, 4, 1996, 363-370.

7.3. *Double-Layer Bar Grids*

52. Diéguez, R.G. and Gómez de Cózar, J.C., "Florín System. Double Layer Spatial Deployable Structures with Frames of Rhombuses and Scissors", Deployable Structures: Theory and Application, Book of Abstracts, IUTAM/IASS Symp., Cambridge, U.K., September, 1998.
53. Fanning, P. and Hollaway, L., "The Deployment Analysis of a Large Space Antena", Int. J. Space Structures, 8, 3, 1993, 209-220.

54. McNulty, O., "Foldable Space Structures", Shells, Membranes and Space Frames, Proc. IASS Symp., Osaka, Sept. 1986, V3, 277-284.
55. Meurant, R., "Circular Space Trusses", Space Structures 4, Proc. 4th Int. conf. Space Structures, University of Surrey, September 1993, V2, 2053-2062.
56. Mitsugi, J. and Yasaka, T., "Deployable Modular Mesh Antena and its Surface Adjustment", Int. J. Space Structures, 8, 1-2, 1993, 53-61.

7.4. *Strut-Cable Systems (DLGs)*

Tensegrity

57. Bouderbala, M. and Motro, R., "Folding Tensegrity Systems", Structural Morphology, Towards the New Millennium, Proc. IASS coll., Univ. of Nottingham, Aug. 1997, 115-122.
58. Hanaor, A., "Double-Layer Tensegrity Grids as Deployable Structures", Int. J. Space Structures, 8, 1-2, 1993, 135-143.
59. Hanaor, A., "Tensegrity: Theory and Application", Beyond the Cube, Gabriel, J.F., ed., John Wiley & Sons, New York, 1997, Ch. 13, 385-408.
60. Le Saux, C., Bouderbala, M., Cevaër, F. and Motro, R., "Strut-Strut Contact in Numerical Modeling of Tensegrity Systems Folding", Shells and Spatial Structures: From Recent Past to the Next Millenium, Proc. IASS 40th Anniversary congress, Madrid, 1999, VII, D1.
61. Motro, R., "Tensegrity Systems and Structural Research", Current and Emerging Technologies of Shell and Spatial Structures, Proc. IASS Col., Madrid, April 1997, 161-173.
62. Motro, R. and Bouderbala, M., "Mobile Tensegrity Systems", Mobile and Rapidly Assembled Structures II, Proc. Intnl Conf. MARAS 96, Computational Mechanics Publications, Southampton, 1996, 103-111.

63. Pedretti, M., "Smart Tensegrity Structures for the Swiss Expo 2001", Lightweight Structures in Architecture, Engineering and Construction, Proc. LSA98, Sydney, 1998, V2, 684-691.
64. Pedretti, M., Internet site http://www.ppeng.ch/base_expo.htm, July, 1998.

Others

65. Kwan, A.S.K. and Pellegrino, S., "A New Concept for Large Deployable Space Frames", Int. J. Space Structures, 9, 3, 1994, 153-162.
66. Kwan, A.S.K. and Pellegrino, S., "Design and Performance of the Octahedral Deployable Space Antena (ODESA)", Int. J. Space Structures, 9, 3, 1994, 163-173.

7.5. *Angulated Scissors (SLG, retractable)*

67. Hoberman, C., Reversibly Expandable Three-Dimensional Structure, U.S. Patent 4,780,344.
68. Hoberman, C., Reversibly Expandable Doubly Curved Truss Structure, U.S. Patent 4,942,700, 1990.
69. Hoberman, C., Radial Expansion/Retraction Truss Structure, U.S. Patent 5,024,031, 1991.
70. Hoberman, C., "Iris Dome", Projects 45, Chuck Hoberman, Exhibition brochure, Museum of Modern Art, New York, February 24-April 12, 1994.
71. Kassabian, P.E., You, Z. and Pellegrino, S., "Retractable Structures Based on Multi-Angulated Elements", Structural Morphology, Towards the New Millennium, Proc. IASS Coll., Univ. of Nottingham, Aug. 1997, 130-135.
72. You, Z. and Pellegrino, S., "New Solutions for foldable Roof Structures", Mobile and Rapidly Assembled Structures II, Proc. Intnl Conf. MARAS 96, Computational Mechanics Publications, Southampton, 1996, 35-44.
73. You, Z. and Pellegrino, S., "Foldable Bar Structures", Int. J. Solids Structures, 34, 15, 1997, 1825-1847.

7.6. *Other Single-Layer Bar Grids*

74. Baverel, O. and Saidani, M., "The Multi-Reciprocal Grid System", IASS Bull., 40, 1999, 33-41.
75. Burry, M. and Popovic, O., "Parametrically Controlled Resolution for Reciprocal Frame Design, Description and Manufacture", Structural Morphology, Towards the New Millennium, Proc. IASS Coll., Univ. of Nottingham, Aug. 1997, 84-91.
76. Chilton, J.C., "Polygonal Living: Some Environment-Friendly Buildings with Reciprocal Frame Roofs", IASS Bull., 36, 2, 1995, 83-89.
77. Chilton, J.C., Choo, B.S. and Wilkinson, D., "A Parametric Analysis of the Geometry of Retractable Reciprocal Frame Structures", Lightweight Structures in Architecture, Engineering and Construction, Proc. LSA98, Sydney, 1998, V1, 547-555.
78. Kent, E.M., "Kinematic Periodic Structures (KPS) based on Revolute Hyperboloid", Mobile and Rapidly Assembled Structures II, Proc. Intnl Conf. MARAS 96, Computational Mechanics Publications, Southampton, 1996, 241-252.
79. Popovic, O., Chilton, J.C. and Choo, B.S., "Rapid Construction of Modular Buildings using the 'Reciprocal Frame'", Mobile and Rapidly Assembled Structures II, Proc. MARAS 96, Computational Mechanics Publications, Southampton, 1996, 73-82.
80. Popovic, O., Chilton, J.C. and Choo, B.S., "The Variety of Reciprocal Frame (RF) Morphology Developed for a Medium Span Assembly Building – Case Study", Structural Morphology, Towards the New Millennium, Proc. IASS Coll., Univ. of Nottingham, Aug. 1997, 164-171.
81. Saidani, M. and Baverel, O., "Retractable Multi-Reciprocal Grid Structures", IASS Bull., 39, 3, 1998, 141-146.
82. Piekarski, M., "Constructional Solutions for Two-Way-Fold-Deployable Space Trusses", Deployable Structures: Theory and Application, Book of Abstracts, IUTAM/IASS Symp., Cambridge, U.K., September, 1998.

7.7. *Spines for Membrane and Hybrid Structures*

83. Adriaenssens, S.M.L., Barnes, M. and Mollaert, M., "Deployable Torsion Free Tensegrity Spines", Engineering a New Architecture, Proc., Aarhus, Denmark, May 1998, pp. 93-101.
84. Burford, N., Fish, D. and Smith, F., "The Development of a Lightweight Military shelter", Transportable Environments, Kronenburg, R., Ed., E&FN Spon, London, 1998, 158-164.
85. Furuya, H., "Concept of Deployable Tensegrity Structures in Space Application", Int. J. Space Structures, 7, 2, 1992, 143-151.
86. Chen, W., Guan, F. and Dong, S., "Development of a Highly Synchronized Deployment Space Mast", J. IASS, 40, 2, 1999, 113-119.
87. Gioncu, V. and Tîrča, L., "Structural Morphology of Hybrid Tensioned Membranes", Structural Morphology, Towards the New Millennium, Proc. IASS Coll., Univ. of Nottingham, Aug. 1997, 324-331.
88. Natori, M. Miura, K. and Furuya, H., "Deployable Truss Concepts in Space Technology", Shells, Membranes and Space Frames, Proc. IASS Symp., Osaka, Sept. 1986, V3, 261-268.
89. Raskin, I. and Roorda, J., "Buckling Force for Deployable Pantographic Columns", Mobile and Rapidly Assembled Structures II, Proc. MARAS 96, Computational Mechanics Publications, Southampton, 1996, 305-314.
90. Raskin, I. and Roorda, J., "Bending of Three Dimensional Pantographic Columns", Deployable Structures: Theory and Application, Book of Abstracts, IUTAM/IASS Symp., Cambridge, U.K., September, 1998.
91. Sastre, R., "Expandable Arches", Mobile and Rapidly Assembled Structures II, Proc. Intnl Conf. MARAS 96, Computational Mechanics Publications, Southampton, 1996, 123-131.
92. Sircovich-Saar, O., "Self-Erecting Two-Layer Steel Prefabricated Arch", Proc. 3rd Int. Conf. Space Structures, University of Surrey, September 1984, Elsevier Applied Science, 823-827.

93. Tsutomu, K. and Tokai, H., "Cable Scissors Arch-Marionettic Structure", Structural Morphology, Towards the New Millennium, Proc. Int. Coll. IASS, Univ. of Nottingham, Aug. 1997, 107-116.

7.8. *Plates*

Folded plates

94. Delarue, J.M., "Minimal Folding Tensegrity Configurations", Structural Morphology, Proc. 1st Intnl. Seminar on Structural Morphology, IASS, Montpellier, September 1992, 101-111 (in French); IASS Bull. 34, 113, 1993, 171-180.
95. Foster, C.G. and Krishnakumar, S., "A Class of Transportable Demountable Structures", Int. J. Space Structures, 2, 3, 1986/87, 129-137.
96. Fuller, R.B., Laminar Geodesic Dome, U.S. Patent 3,203,144.
97. Guest, S.D. and Pellegrino, S., "Inextensional wrapping of Flat Membranes", 1st Intnl. Seminar on Structural Morphology, IASS, Montpellier, September 1992, 203-215.
98. Sedlak, V., "Paperboard Structures", Proc. 2nd Intnl. Conf. Space Structures, University of Surrey, Guildford, September 1975, 780-793.
99. Specht, B., "Deployment Simulation of a Parabolic Space Antenna with Rigid Panels", Int. J. Space Structures, 8, 1&2, 1993, 63-70.
100. Tonon, O.L., "Geometry of Spatial Folded Forms", Int. J. Space Structures, 6, 3, 1991, 227-240.
101. Tonon, O.L., "Geometry of the Spatial Folded Forms", Space Structures 4, Proc. 4th Int. conf. Space Structures, University of Surrey, Guildford, September 1993, V2, 2042-2052.

Curved surfaces

102. Bini, D.N. and Pietrogrande, S., "Self-Shaping Space Structures", Int. J. Space Structures, 4, 2, 1989, 67-80.
103. Hansen, K.F., "A Method for Faceting Double Curved Surfaces", Structural Morphology, Proc. 1st Intnl. Seminar on Structural

Morphology, IASS, Montpellier, September 1992, 169-178; IASS Bull. V34, N113, 1993, 196-200.
104. Huybers, P., "The Polyhdral World", Beyond the Cube, Gabriel, J.F., ed., John Wiley & Sons, New York, 1997, 243-279.
105. Pavlov, G.N., "Determination of Parameters of Crystal Latticed Surfaces Composed of Hexagonal Plane Facets", Intnl. J. Space Structures, 5, 3&4, 1990, 169-185.
106. Sogame, A. and Furuya, H., "Conceptual Study on Cylindrical Deployable Space Structures", Deployable Structures: Theory and Applications, Book of Abstracts, IUTAM/IASS symp. Cambridge University, September, 1998.
107. Wester, T., "Plate-Domes", Proc. Intnl. conf. on the Design and Construction of Non-conventional Structures, Topping B.H.V., ed, Civil Comp Press, London, 1987, V1, 241-250.
108. Wester, T., "A geodesic Dome-Type Based on Pure Plate Action", Intnl. J. Space Structures, 5, 3&4, 1990, 155-167.

7.9. *Fabric Structures*

109. Hooper, P., "Tensile Tiles: The Cubic Modularity of Curved Surfaces", Lightweight Structures in Architecture Engineering and Construction, Proc. LSA98, LSAA, Sydney, 1998, V2, 1007-1018.
110. Karni, E., "Retractable Spatial Structures for Swimming Pool Enclosures", Int. J. Space Structures, 10, 4, 1995, 231-236.
111. Kronenburg, R., Future Tents Limited (FTL), Softness, Movement and Light, Academy Editions, London, John Wiley, New York, 1997.
112. Mollaert, M., "Retractable Membrane Roofs", Mobile and Rapidly Assembled Structures II, Proc. MARAS 96, Computational Mechanics Publications, Southampton, 1996, 407-417.
113. Mollaert, M., Hebbelinck, S. and Adriaenssens, S., "Modular Membrane elements for Dismountable Structures", Lightweight Structures in Architecture Engineering and Construction, Proc. LSA98, LSAA, Sydney, 1998, V2, 760-767.

114. Saitoh, M., "Recent Developments of Tension structures – Aesthetics and Technology of the Strings", Current and Emerging Technologies of Shell and Spatial Structures, Proc. IASS Coll., Madrid, April 1997, 105-118.

7.10. *Pneumatic Structures*

115. Herzog, T., Pneumatic Structures, A Handbook for the Architect and Engineer, Crosby Lockwood Staples, London, 1977.
116. IASS W.G. 7, Recommendations for Air-Supported Structures, IASS, Madrid, 1985.
117. IL 15, Lufthallenhandbuch – Air Hall Hanbook, Institute for Lightweight Structures (IL), Stuttgart, 1982.
118. Kim, J.Y., Plaut, R.H. and Kwun, T.J., "Analysis of Tent-Like Structures Supported by Pressurized Fabric Arches", Lightweight Structures in Architecture Engineering and Construction, Proc. LSA98, Sydney, 1998, V2, 821-829.
119. Kröplin, B. and Wagner, R., "Pneumatic Tube-Structures, Some Examples", IASS Bull., 36, 2, August, 1995, 67-72.
120. Riches, C.G. and Gosling, P.D., "Numerical Analysis of the Load-Deflection Characteristics of Pneumatic Beam Structures", Lightweight Structures in Architecture Engineering and Construction, Proc. LSA98, LSAA, Sydney, 1998, V2, 875-882.
121. Riches, C.G. and Gosling, P.D., "Pneumatic Structures: A Review of concepts, Applications and Analytical Methods", Lightweight Structures in Architecture Engineering and Construction, Proc. LSA98, LSAA, Sydney, 1998, V2, 883-894.
122. Szyszkowski, W. and Glockner, P.G., "The Use of Membrane Structures in Space", Int. J. Space Structures, 5, 2, 1990, 106-129.
123. Trushchov, A.G. and Ivanov, G.P., "The Use of Pneumatic Rings and Spirals in Space Structures and Framework Systems", Int. J. Space Structures, 9, 4, 1994, 211-217.

CHAPTER 7

PHANTASY IN SPACE: ON HUMAN FEELING BETWEEN THE SHAPES OF THE WORLD AND HOW TO LOOK ON NATURAL STRUCTURES

Michael Balz

Dipl.-Ing. (SBS), Stuttgart, Germany

mechanism of swelling is much more complicated and is not understood. In order to have a better insight into the problem of the swollen form of viruses, we started with the investigation of the motion of a simpler object: an expandable dodecahedron.

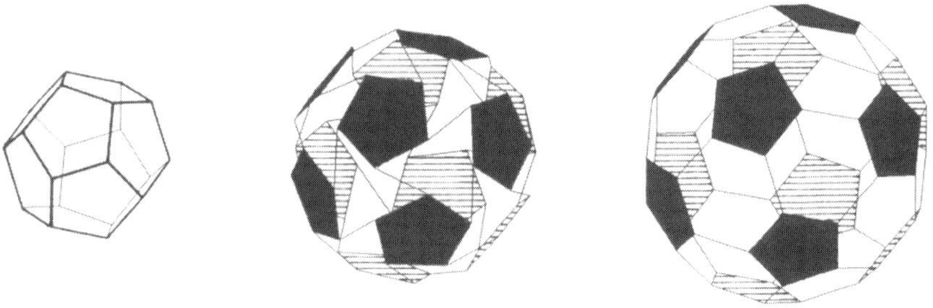

Fig. 1. An expandable dodecahedron.

In the research of fullerene chemistry, the leapfrog transformation is known [Ref. 1]. If the polyhedron of the original fullerene is capped on every face, a deltahedron is obtained. If this deltahedron is then converted to its dual, a new fullerene polyhedron with three times as many vertices is obtained. This procedure called leapfrog is a discontinuous transformation because it jumps one fullerene to another over the intervening deltahedron. The leapfrog transformation converts a dodecahedron to a truncated icosahedron. It turns out that the dodecahedron → truncated icosahedron leapfrog transformation can be described as a continuous process using Verheyen's dipolygonid [Ref. 7] composed of 30 digons and 12 pentagons, denoted by $30\{2\} + 12\{5\} \,|\, 31.717474°$.

We have made a cardboard model that expands from a dodecahedron to a truncated icosahedron by applying a rotation on one pentagon while keeping the opposite pentagon fixed (Fig. 1). The model represents a mechanism composed of rigid bodies with hinged connection. It has more infinitesimal degrees of freedom; one of them is responsible for swelling. In this paper we investigate the motion of our expanding structure.

2. Physical Model

Both expandable polyhedral viruses and fullerene molecules share the basic property that all pentagons (or hexagons) are situated symmetrically at any position. Therefore, our first aim was to find a mechanism producing this symmetrical motion.

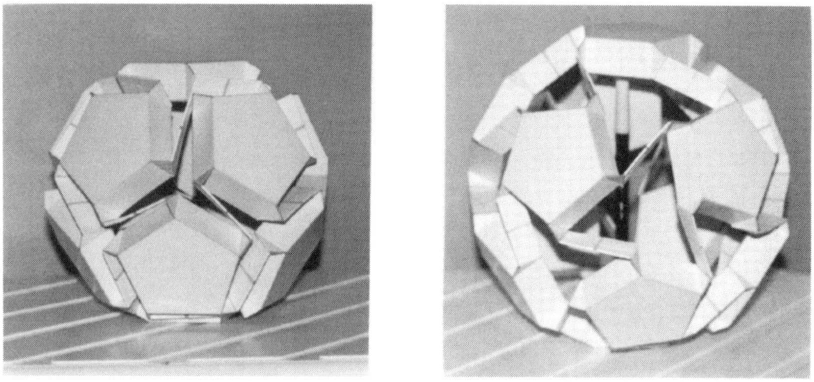

Fig. 2. A cardboard model of the expandable dodecahedron.

2.1. *The Structure*

Our cardboard model (Fig. 2) was designed upon the following principles:

(i) the structure may only consist of rigid bodies (pentagons) and connecting elements (bars, plates),
(ii) all conditions of compatibility should be satisfied in any position while expanding,
(iii) all joints in the structure are considered to be hinges or ball-joints (no sliding connections should be applied),
(iv) all connecting elements (or structures, if the connection itself contains more rigid elements) between adjacent pentagons are to be identical and have to have a two-fold rotational symmetry (C_2).

Note that principle (iv) is a necessary but not sufficient condition in order that two-fold rotational symmetry between adjacent elements (and the symmetrical behavior of the whole structure as well) is preserved, since two connected pentagonal members may incline to the same connecting element under different angles even if it is symmetrical.

In this particular model pentagon-based prisms and planar connecting elements (triangles and rectangular quadrangles) were applied, the connection between them was made by hinges in any case. In Fig. 3 two prisms and their connection are shown.

2.2. *Type and Range of Expansion*

A clearly detectable motion of the structure – according to the main requirement – is a symmetrical expansion. It means that each pentagon rotates about its axis of symmetry by the same angle and moves radially outwards with the same speed – consequently, any position of a pentagon is parallel to its initial one. In other words, all inclination angles between adjacent pentagons remain constant – let it be denoted by i, so $\angle ABC$ and $\angle DEF$ are equal to $i/2$ in Fig. 3.

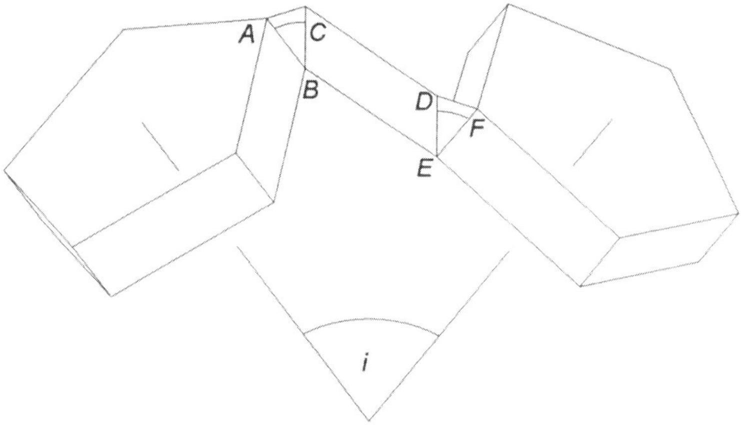

Fig. 3. Two pentagonal prisms and their connection.

If any of the pentagons appears to have two different (but not independent) motions, a geometrical relationship can be determined between them. For an easier discussion, let us consider a simplified network of the model where only the inner vertices of the pentagonal prisms are taken into account (Fig. 4). Let the edge length be denoted by a. Since the rotation of two adjacent pentagons about their own axis requires the rotation of the linking element (digon) as well and the digons also rotate about their radial symmetry axis, we can define – for the sake of simplicity – this angle of rotation (α) as a function of the radius (R) of the circumscribed sphere.

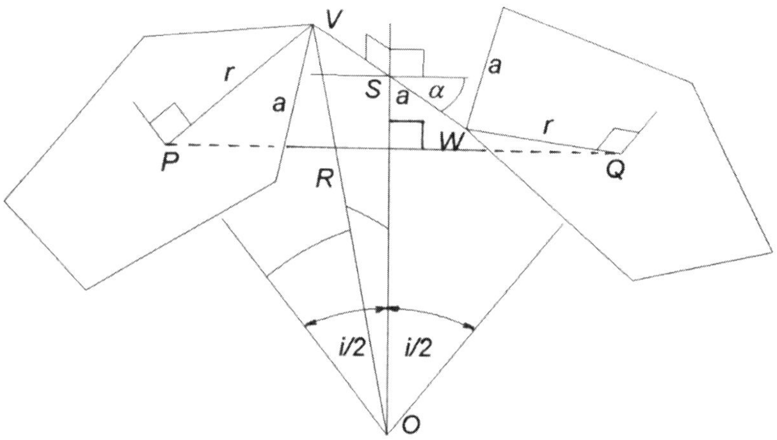

Fig. 4. Simplified network of two pentagonal members (inner edges and vertices only).

Let us project the mid-points of three mutually adjacent faces of a dodecahedron to the circumscribed sphere (points X, Y, Z, Fig. 5).

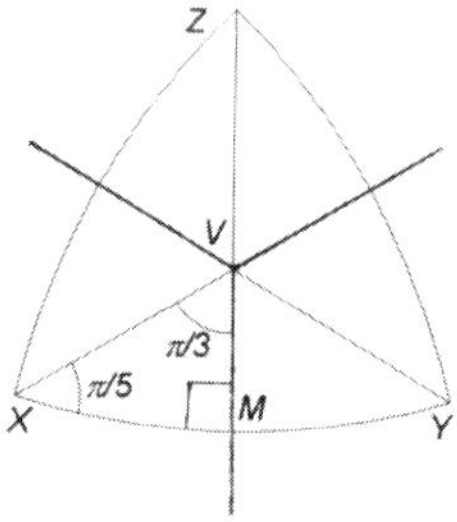

Fig. 5. A vertex with 3 adjacent faces (spherical projection).

Now the arc drawn from the vertex V to the mid-point of arc XY (M) is perpendicular to arc XY. The cosine theorem for the spherical triangle VMX shows that:

$$\cos\angle XVM = -\cos\angle VMX \cos\angle MXV + \sin\angle VMX \sin\angle MXV \cos XM\,,$$

where $XM = i/2$, $\angle XVM = \pi/3$, $\angle VMX = \pi/2$, $\angle MXV = \pi/5$.

From this equation

$$\cos\frac{i}{2} = \frac{\cos\dfrac{\pi}{3}}{\sin\dfrac{\pi}{5}} = \frac{1}{2\sin\dfrac{\pi}{5}}$$

is obtained.

Looking at Fig. 4 again, another formula of the same theorem for triangle PSV says:

$$\cos VP = \cos PS \cos SV + \sin PS \sin SV \cos\angle PSV\,,$$

where $\sin VP = \dfrac{\overline{PV}}{\overline{VO}} = \dfrac{r}{R} = \dfrac{a}{2\sin\dfrac{\pi}{5}R}$, $PS = i/2$,

$\sin SV = \dfrac{a}{2R}$ and $\angle PSV = \alpha$.

From this formula, $\cos\alpha$ can be expressed:

$$\cos\alpha = \frac{\cos VP - \cos SV \cos PS}{\sin SV \sin PS}\,,$$

substituting $\sqrt{\dfrac{5-\sqrt{5}}{8}}$ for $\sin\dfrac{\pi}{5}$ we obtain that

$$\cos\alpha = \frac{2\sqrt{\dfrac{5-\sqrt{5}}{2}\left(\dfrac{R}{a}\right)^2 - 1} - \sqrt{4\left(\dfrac{R}{a}\right)^2 - 1}}{\sqrt{\dfrac{3-\sqrt{5}}{2}}} \qquad (1)$$

and its inverse function (2):

$$\frac{R}{a} = \frac{\sqrt{\left(4+\sqrt{5}\right)\cos^2\alpha + 2\cos\alpha\sqrt{\left(5+2\sqrt{5}\right)\cos^2\alpha + 9 + 4\sqrt{5}} + \left(9+3\sqrt{5}\right)/2}}{2}$$

Since $\alpha = \pi/2$ means the most compact (dodecahedral) configuration ($R = R_{min}$) while $\alpha = 0$ the fully expanded (truncated icosahedral) one ($R = R_{max}$), we can calculate the minimum and maximum values for R/a:

$$\frac{R_{min}}{a} \approx 1.401,$$

$$\frac{R_{max}}{a} \approx 2.478.$$

The quotient of these values shows that the circumradius of our model can increase by about 77% during the expansion.

As to the volumes of the polyhedra, another calculation can be carried out. The volume of the dodecahedron is $a^3\sqrt{\frac{235+105\sqrt{5}}{8}} \approx 7.6631a^3$, while the same number for the truncated icosahedron is $a^3\left(\frac{45}{4}\left(3+\sqrt{5}\right) - \sqrt{\frac{15+5\sqrt{5}}{2}}\right) \approx 55.2877a^3$. Consequently, the volume of the expanded structure is 7.215 times larger than that of the dodecahedron.

3. Numerical Models

In the previous chapter we analysed a special motion of the expandable dodecahedron. It is still not known, however, whether the structure is able to move only in that way or it has other independent motions. In order to answer this question, it is necessary to execute a detailed analysis of the compatibility matrix of the structure.

3.1. *A Bar-and-Joint Structure*

The simplest way to investigate the kinematical behaviour of the system is to consider a pure bar-and-joint assembly. Let the compatibility (or Jacobian) matrix be denoted by **J**, the number of its columns by c and the number of rows by r. In this case, c will be equal to the triple of the number of free nodes (n) in 3D, r will give the number of bars in the structure.

In Fig. 6 a possible network is shown that is sufficient to ensure the rigidity of both the prism and the quadrangular element of the connection. It is known that, in a generic case, a polyhedron with v vertices can be made rigid by applying $3(v\text{-}2)$ bars for bracing – this means $3(10\text{-}2) = 24$ for each prism. A planar quadrangle, however, cannot be considered as a generic arrangement of the four nodes: an extra non-coplanar vertex is needed to exclude the infinitesimal motion that will require $3(5\text{-}2) = 9$ bars for each quadrangle. The remaining triangles can now be represented by only one bar each (e.g. bar AC in Fig. 6).

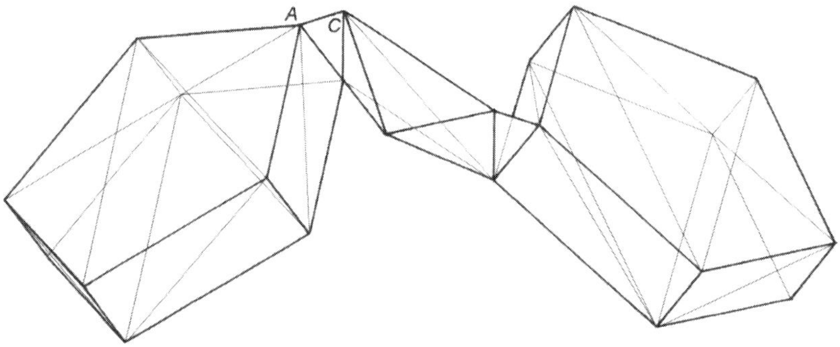

Fig. 6. Network of the bar-and-joint structure.

As to the number of free nodes, we need to take into consideration on one hand that two of the five nodes of the 'digon' coincide with one node of the adjacent prisms; on the other hand, it seems to be useful to avoid rigid body motions, so we imagine one of the prisms fixed. It means that ten nodes are excluded from the free ones, consequently, the total number of free nodes will be equal to:

$$n = (12\text{-}1) \times 10 + 30 \times (5\text{-}2) = 200$$

and

$$c = 3n = 600.$$

Under the same conditions, r can also be determined as follows:

$$r = (12\text{-}1) \times 24 + 30 \times 9 + (2 \times 30) \times 1 = 594.$$

Since c represents the number of unknowns, r that of the conditions, we have got an important information about the kinematical behaviour, namely, $c - r = 6$ means six independent infinitesimal motions for the structure.

This calculation, however, still does not give the final answer for the question that how many independent motions the structure has – it can only be decided by investigating the rank deficiency of **J**. It implies another problem, because the computational process (especially if we want to get information about the characteristics of the detected motions) is very sensitive to the size of the matrix. It was this recognition that led us to a new approach of numerical modelling.

3.2. *An Alternative Model*

In our previous model there was only one type of unknowns (the nodal coordinates). At the same time it was necessary to use only one type of constraint functions that expressed the difference between actual and original (l_{ij}) length of a bar P_iP_j:

$$F^b{}_{ij} = \sqrt{(x_j - x_i)^2 + (y_j - y_i)^2 + (z_j - z_i)^2} - l_{ij} = 0 \tag{3}$$

The derivative of $F^b{}_{ij}$ according to any coordinate gives the entry of **J** in the column corresponding to the coordinate and in the row corresponding to the bar P_iP_j. For example:

(iii) The previous statement is true only if the length of $\mathbf{v}_l$ does not vary. If the original length was v_l then:

$$F'_l = \sqrt{x_l^2 + y_l^2 + z_l^2} - v_l = 0\,. \tag{7}$$

In this new model c will involve not only nodes but vector coordinates as well. Keeping a supposition that one of the pentagons is fixed and calculating with 7 bars per pentagon and one per digon the following values are obtained for c and r (the terms are in the order of constraint functions):

$$c = 3 \times ((12\text{-}1) \times 5 + 30 \times 1) = 255,$$

$$r = ((12\text{-}1) \times 7 + 30 \times 1) + (12\text{-}1) \times 2 + 30 \times 1 + 2 \times 30 \times 1 + 30 \times 1 = 249.$$

It is seen that $c - r = 6$ is unchanged, so this model gives the same result with a less than half-size matrix.

4. Analogies and Results

It is possible to construct similar assemblies not only for dodecahedron but for an arbitrary polyhedron (some angles and lengths may be changed but not the topology of the system). Note that nothing can be stated about the movability of these structures in general. Let us denote the number of faces, edges and vertices of the base polyhedron by F, E and V respectively. If the number of edges intersecting at one vertex of the polyhedron is the same for each vertex, let then this number be Q.

For this case the following equation holds:

$$(VQ)/2 = E. \tag{8}$$

In the followings we will determine values c-r for these structures based on 'Q-valent' polyhedra.

Our models can be imagined as assemblies of rigid bodies and hinges, exclusively. It is known that a rigid body in space has 6 degrees of freedom of motion, while a hinge is a connection that means a 5-degree constraint, allowing for only one motion. The connecting structure between two polygon-based prisms can be decomposed into three rigid bodies (two triangles and a quadrangle) and four hinges. If we count all the rigid bodies (B) and hinges (H) in a structure (the fixed prism is still excluded), we get the following values:

$$B = (F\text{-}1) + 3E,$$

$$H = 4E.$$

Now calculating the value of c-r the result is:

$$c - r = 6B - 5H = 6(F\text{-}1) - 2E. \tag{9}$$

From the Euler polyhedron theorem it is known that

$$E + 2 - V = F. \tag{10}$$

Substituting (10), then (8) into (9) we obtain

$$c - r = 2VQ - 6V + 6. \tag{11}$$

If the polyhedron is trivalent (Q = 3), the right-hand side of (11) becomes equal to 6. It can be stated that any of these structures based on a trivalent polyhedron has six columns more than rows in its compatibility matrix regardless of the number of faces, edges or vertices.

This theorem can also be used for the analysis of a case analogous to that of the dodecahedral model. Among the five regular polyhedra there are three trivalent ones: the dodecahedron, the cube and the tetrahedron. This fact permits the supposition that an expandable tetrahedral model shows characteristics similar to those of a dodecahedral one, and because the first case needs much less computation, its analysis has been executed by using singular value decomposition of **J** [Ref. 5].

Following the second numerical method, 9 free nodes and 6 free vectors were taken into account, while the number of constraints came from the sum $r = ((4\text{-}1) \times 3 + 6) + 0 + 6 + 2 \times 6 + 6$, that is, $c = 45$ and $r = 39$. The computational analysis gave the result that the compatibility matrix had a single rank deficiency, in other words: the structure had have to be completely newly created with the aid of model measurements and reengineering in the computer.

Here the two main beams of the structure - completely independent from the shell - were always meant to be of strongest and genuine expression for the main assembly hall in the building. Already their preliminary form, approximated in a couple of models, showed their specific properties: continuously changing instantaneous radius, strong occurrence of torque. Due to the complexity of their shape their models in a scale of 1:25 came out rather poorly. Additionally they could not be photographed adequately by stereo photogrammetry. An attempt to measure them in a 3-D mechanical measurement device met similar difficulties.

Thus only a limited number of coordinates of significant points on the surfaces and edges could be obtained, supposedly not very exact and moreover liable to errors for lack of identification, moreover, due to the unfavorable position of this edge-line with respect to the position of the camera, rather large errors in the measured co-ordinates were to be expected.

But even if one had succeeded in measuring the models most exactly this would not have been sufficient to base premanufacturing on the measurements only:

- Contrary to their representation in the model the real shells were reshaped at their mutual joint lines and the main beams given special significance by "extracting them" from the continuous surface of the shells to become independent primary architectural elements.
- The vertical beams ends were to contact directly in thin seams the vertical walls of the internal large assembly hall, constructed of bricks. This constraint had not been realized in the model.
- The two main beams are distinctive elements of expressive shapes. As such they had been sketched, experimented with in models, and thoroughly discussed between client, architect and engineers during a long time. However the client could not - and this is quite natural - express his vision using a language of geometry and mathematics. However, his vision was by no means blurred and foggy but vivid and clear. It took considerable time until the engineers could emphatically sense what the client had in mind.

- Also the glulam premanufacturer issued further constraints defining minimum radius of curvature and torque to be adhered to.

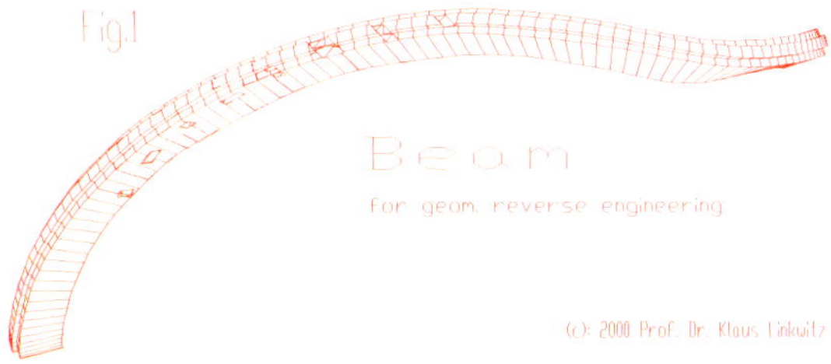

Fig. 1. Geometrical reverse engineering.

As a consequence the main beams had to fully redesigned using geometrical reverse engineering, using as point of departure models, measurements, and discussion.

Starting from here a considerable number of geometrical/ mathematical experiments were undertaken in the computer and the intermediate results in due course shown to the client asking his judgment and comments. Progressively it became possible to correlate the vision of the client with the parameters of the mathematical tools and to improve the redesign step by step with an ever steepening gradient, achieving finally congruence of imagination and computer model.

The shape finally adopted then had to be translated, as is usual, into digital numbers to monitor the manufacturing machinery of the glulam factory. In this process it became also necessary to subdivide each main beam into five separate pieces, constituting the assembly units to be joined on the building site with wedge-joints.

2. Geometrical/Mechanical Constraints Imposed on Design to make Execution Feasible and Subsequent Reverse Engineering. "Build and Push Method" (B&P-Method) of Bridge Construction: Geometrical Reverse Engineering Applied to the Loisach Bridge

The "Build and Push Method" (B&P – Method)

The Build and Push Method originates from Stuttgart. Invented by the partner *Bauer* of the consulting engineering firm *Leonhardt & Andrä,*

Stuttgart in the sixties, it was applied for the first time for the construction of the *Caroni-River-Bridge* in Venezuela. Stimulating the introduction of the new method was the fact, that the Caroni-River varies about 10m in height between low and high waters, has a swift currency and is notoriously susceptible to quick changes in water level caused by unstable weather condition in its region of inflowing rivers. Thus the erection of a strong and reliable scaffolding for the construction of the bridge body would have been complicated and expensive.

The - at that time revolutionary - -idea of *Bauer* was, to have the whole bridge body of about 200 m length constructed off the river on one of the banks; then lying there on rails after termination of these partial works. Simultaneously, in the river, the pillars of the bridge were to be constructed. Then, in the final step of construction, the bridge was to be pushed from the bank over the pillars with the aid of hydraulic pressure gadgets, gliding on its path from the original construction site to its final position, first on the rails and then on the pillars. To lower friction during this process the pillar bearings would be coated by *Teflon*, also newly invented at that time.

The CEO *Lenz* of the executing Company *Züblin*, also from Stuttgart, highly attached to new progressive if economic methods of execution and not afraid of calculated risks, and after promising to nail and hang *Fritz Leonhardt*, Senior CEO of *Leonhardt & Andrä* to the first pillar if the construction should fail, got the contract and succeeded with execution.

Ever since then bridges have been constructed using the B&P-Method. First only bridge designs of simple geometries were considered for pushing: Elevation and plan of the bridge are straight lines. Then combinations of straight lines and circles in elevation and plan were tried and successfully executed. An attempt in Switzerland of building and pushing a bridge the geometry of which consisted of both circles in elevation and plan failed. During midway construction, the bridge being pushed progressively encountered ever increasing friction on the pillars and finally seized to a standstill like the piston in a cylinder when oil is drained out.

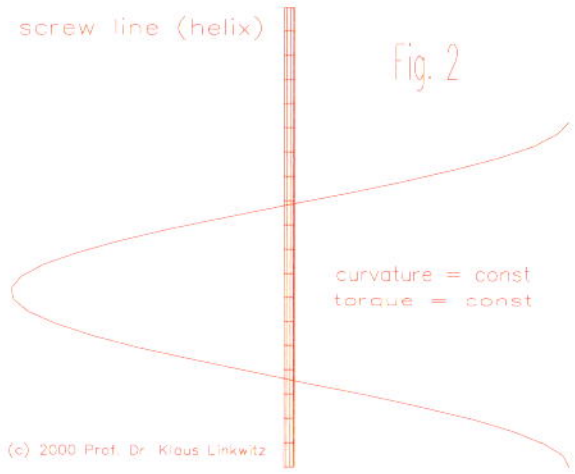

Fig. 2. Single screwline, elevation.

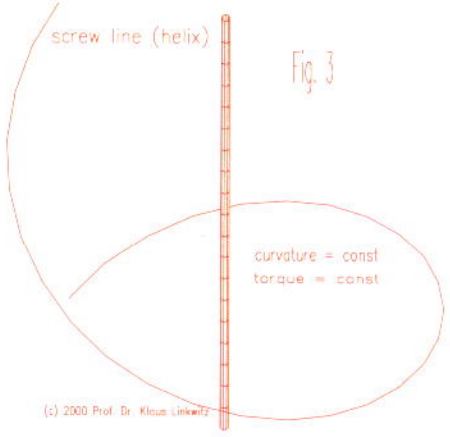

Fig. 3. Screw line, parall. Proj.

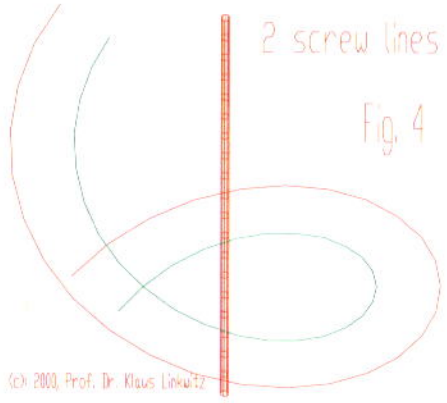

Fig. 4. Double Screw line.

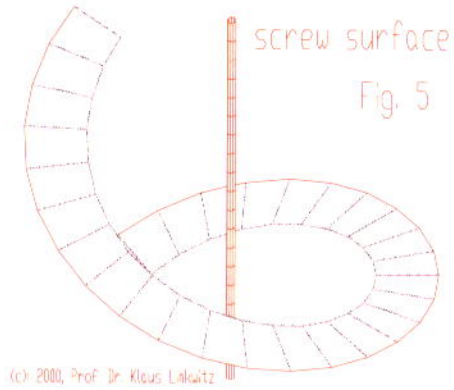

Fig. 5. Screw surface.

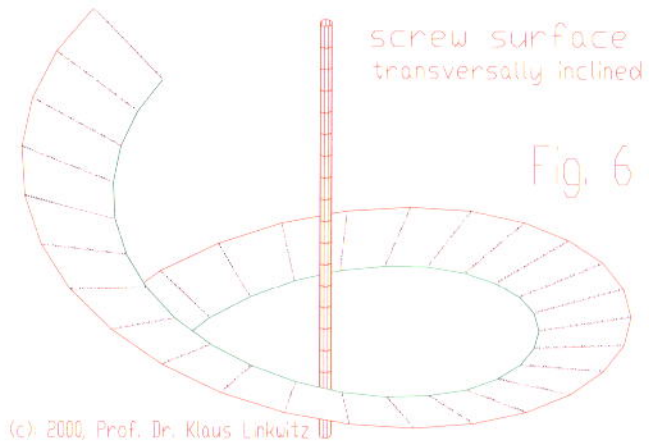

Fig. 6. Transversal inclination of surface.

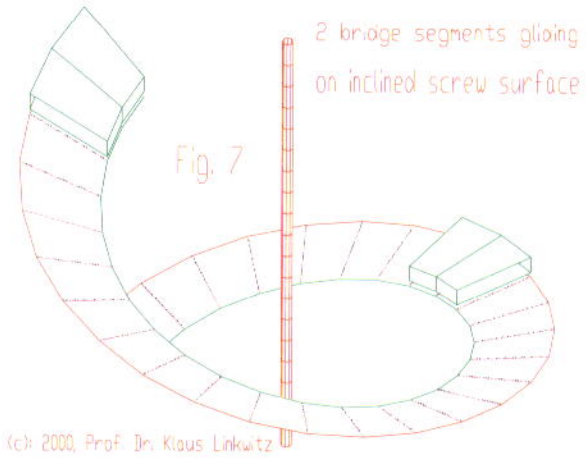

Fig. 7. Bridge segments gliding.

If we want to check, whether the design of a bridge can be constructed by the P&B-Method, i.e. whether the design is "pushable", we can make a mental experiment. We imagine instead of the actual small number of pillars supporting the bridge now a rather indefinite number of such pillars, forming consequently a continuous track upon which the bridge will glide when pushed. This gliding track would reach from either abutment to the middle of the bridge if we would construct the bridge in two halves and push it then from both sides, or it would reach from the abutment on the one side of the river/valley to the abutment on the other side of the river/valley if the bridge were constructed in one piece and then pushed from abutment to abutment.

During pushing each bridge segment has to glide from its original construction site to its final position of the bridge in situ. (Fig. 7) Imagining the most favorable case of a bridge being prefabricated in two pieces and then pushed from the abutments on the banks on either side, the front segment of the bridge has to glide from the abutment unto the middle of the bridge, the second segment has to glide until it reaches its location behind the middle segment, and so on. As each segment has to glide smoothly it has to fit perfectly into the form of the gliding track - i.e. the bearings on the pillars - in its transition from site of premanufacturing to its final location in the bridge.

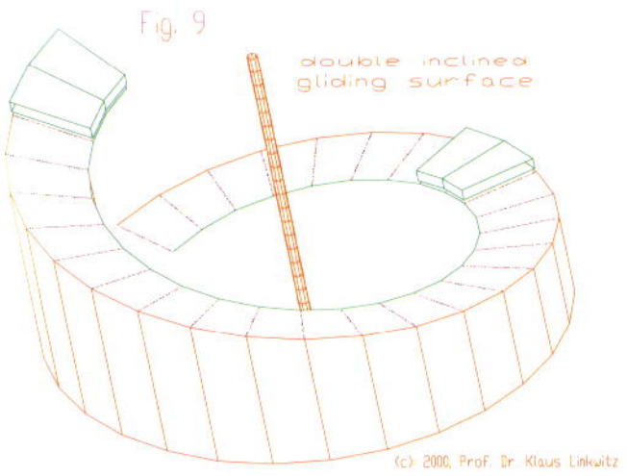

(8)

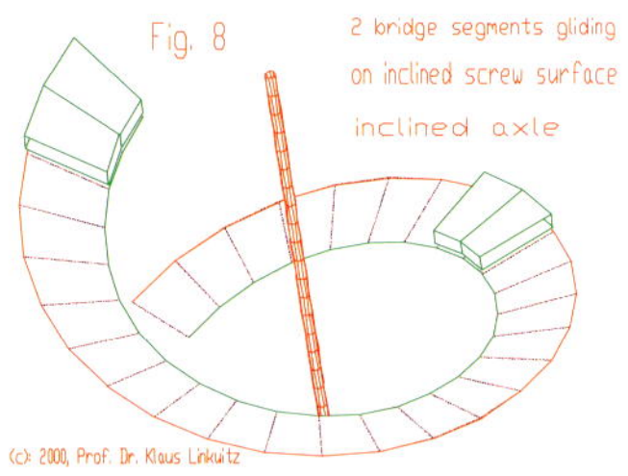

(9)

Figs. 8 and 9. Inclination of axis as a further means of fit.

As a result of our mental experiment we conclude, that the geometry of all gliding parts of the bridge has to be exactly identical with the geometry of the gliding track, materialized by the discrete locations of the pillars. As a consequence - this is noted here without proof – “pushing lines” and “pushable designs” must be - regarded in 3-dimensional space! - straight lines, circles or screwlines (Fig. 2), and the appertaining surfaces. Hence, for example, a design consisting of a combination of circles in plan and circles in elevation generally is not pushable.

In the case of the *Loisach-Bridge*, permitting the highway crossing of the river Loisach near the Alps in Southern Bavaria, one of the competing construction companies had been awarded the contract based upon its alternative design proposing the bridge to be constructed by the P&B-method and consequently was the cheapest among the competing bids. However, the contract of execution had been allotted by the client without a thorough control, whether the already finalized engineering design was pushable or not. To quieten his conscious and to support his ideas the contractor had made a model of bridge and pushing track. Experiments with these seemed to support feasibility of the method and the client could be convinced to accept the alternative design.

Thus the first question to be answered was "Is this design - consisting essentially of compound circle-like curves - pushable or not?"

As a first answer, a mechanical model was made once again, this time with more consequent rigor, by sawing in elevation with a circular seam through a box like trough shaped in plan as a perfect circle, yielding after the cut two halves of the trough. Consequent experiments with these two halves demonstrated clearly, that they would fit only in their final position but could not slide over each other without gaps at the seam. Obviously, the design, as it was, could not be executed by the P&B-Method!

Geometrical Reverse Engineering of essential parts of the design

Here the reverse engineering consisted of three essential parts, namely

(i) Geometrical analysis of the finalized engineering design; especially with respect to curvature and torque of all lines/surfaces which have to be glided
(ii) Computer aided mathematical generation of feasible sliding curves/surfaces and approximation and reshaping of all relevant gliding parts of the finalized engineering design using slidable curves/surfaces
(iii) Analytical modeling and simulation of all stages of the pushing process with the reengineered shape and control of the achieved fit of the bridge on the (also redesigned) pillar bearings.

The drawings for the design of execution comprehended also an exhaustive and complete geometrical description of the bridge i.e. in plan and longitudinal section the radius of all circles, parameters of the transition curves, lengths of all successive curves; in the cross sections the "twisting movements" of the roadway as a function of the radius of the curves in plan, including descriptions of the rotation centers of these movements, etc.

Using differential geometry - extended to discrete difference geometry - in analyzing the contents of the plan and descriptions the actual instantaneous 3D-curvature and -torque of the design's gliding lines and surfaces at densely spaced discrete points could be determined. The spatial radius varied between 1795 and 1805 m and radius of spatial torque between 80 000 and 100 000 meters, indicating clearly that the actual design was not pushable. However, the variations of both radius were so small - compared to the dimensions of the bridge - that approximating screw-lines could be calculated, to be fitted with the aid of least square techniques to the design.

However, as is very obvious, actual gliding is not performed on two-dimensional lines in space but on three-dimensional surfaces. Thus every two calculated glidable screw-lines had to be put together to screw-surfaces. In the assembly of every two gliding lines into one gliding surface also the lateral rotation of the roadway in its longitudinal development had to be accommodated. This however could only be achieved by again slightly distorting the theoretical perfect gliding lines. As a result, also after reengineering the bridge had no perfect pushing shape. Comparing ideal lines and surfaces with the consequent reshaping of relevant parts of the bridge, small gaps between bridge and pillars during the process of pushing were unavoidable resulting in frictions and elastic deformations to be expected when the bridge would be pushed.

To know the frictions to be expected and to estimate the elastic deformations necessary to slide the trough of the bridge, about 30 different progress stages of the pushing procedure were calculated in advance and the gaps, then to be compensated by elasticity, determined. Altogether it became possible to approximate the ideal pushable lines and surfaces so closely by the geometrically reversed design that the remaining deviations were so small as to become nearly unnoticeable on eyesight and the remaining constraints could be overcome by elastic deformations as a new structural analysis showed.

3. Shape of an Existing Structure as Essential Constraint for Subsequent Reverse Engineering: Geometrical Reverse Engineering at Recent Optimisation Works at the Timber Shell Bad Dürrheim

Timber shells and building physics

When the interior of timber shells is permanently exposed to humidity, the efficiency of the insulation and the reliability of the vapor barrier are especially important. Every incident of leakage permitting the humid and warm air to escape from the bathing environment in a non controlled manner would not only induce high costs of energy losses but also have the dangerous potential of generating condensing humidity at the places of escape causing unpredictable interactions with the surrounding timber.

In the example outlined, after more then 10 years of use, the daily operation of the spa covered by the shell was to be improved economically by energy optimizing measures acting simultaneously as a precautionary measure against hidden leakages: The crucial upper parts of the glass-facade and its adjoining edge-beams and ribs become object of sanitation measures by which especially the edge beam was to be removed completely and replaced by another one, reinforced simultaneously by unto date means of insulation.

The new, improved edge beams as objects of geometrical reverse engineering.

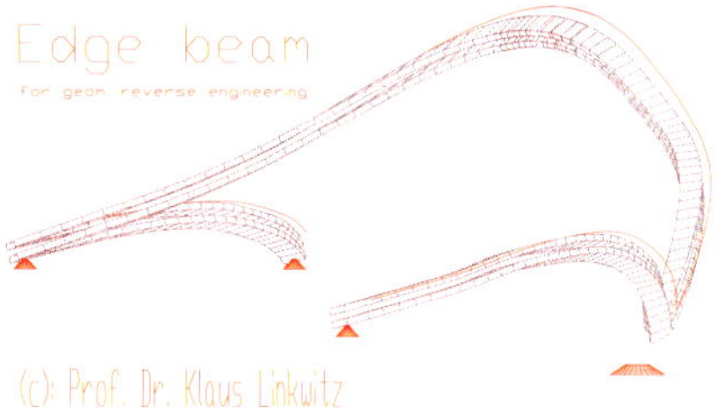

Fig. 10. Four out of the total of seventeen edgebeams.

For the optimization process, the existing edge beam consisting of seventeen individual double beams with dimensions 0,29 x 1,20 m in cross sections and a total length of above 200 m, was completely to be removed, taken out of use, and then replaced piece by piece. This had to be performed under the constraint, that the use of the spa inside must continue all the time. Consequently a building procedure was adopted of working at the same time only at one or two of the total of 17 edge-beams. The selected beams for renewal were then subjected to the most delicate and sensitive work of vertically cutting them with a compass saw in the night and covering the resulting slot with carefully prepared planks immediately. Indeed almost nobody of the spa's guests realised the ongoing very difficult and most professionally performed works undertaken with much prevision and observing all means of security possible.

Then the newly pre manufactured edge beams were to be put into the slot, connected to the insulation, laid on their bearings and connected at their abutments. This is a most sensitive and delicate work. Due to double curvature of the old, removed edge beam, naturally also the slot created by their removal has double curvature everywhere. The slot has not only to accommodate the insulation layers but also the freshly pre manufactured new edge beams, supposed to fit exactly into the gap. Due to their voluminous dimensions, their weight, and the very limited space of manoeuvring practically no corrections on site were possible.

Due to the new insulation and an overall optimization the position and shape of old and new edge beams are *distinctively different* and the new pre manufacturing could *not* use the old manufacturing data. Moreover, all computations performed for the original design and pre manufacturing about 15 years ago had been performed on mainframes no longer existing and the old data were either destroyed or no more readable.

Under these circumstances the only possibility of performing the work was that of a complete fresh creation of the edge beams using the technique of geometrical reverse engineering.

This was achieved by starting from the old shape of the shell in the immediate neighborhood of the edge beams and create from there - using their new design dimensions - new edge beams. Due to the strong double curvature of the shell enough points of identification in unequivocal manner could be found which were usable to fit the edge beams mathematically into the shell. Interpreted as a fitting problem this is highly redundant because the fitting has to be performed following the continuous common lines of beams and the shell.

The solution, based on the methods of least squares, was newly conceived from scratch, translated into appertaining algorithms and implemented in a new computer program consisting of many modules. Thus it became possible to calculate anew all necessary data for the digital monitoring of pre manufacturing with high accuracy and reliability.

Another difficult detail was the redesign of the bearings and metal junction pieces at 17 bearing points, one at either side of each edge beam, as also location and type of all abutments had changed. This was further complicated by the fact, that the actual locations of the constructed bearing differed slightly - but still too much to be neglected - from their design locations shown in the original execution drawings. These differences were determined on spot by measurements and then included into the geometrical reverse engineering process, resulting finally in new drawings and consequent new manufacturing of all bearings.

The challenging works outlined here and demonstrated in the oral presentation with a number of sketches and slides demanded a close co-operation between the glulam company, the principal engineers involved, and the author responsible for providing the necessary computational reconstructions and the digital instructions for monitoring the pre manufacturing. The works were successfully terminated about a year ago without any accident or misfitting.

4. Conclusion

Geometrical reverse engineering is the adequate method when the essential concept of a design and its geometry can only be materialized through a model, or when other complex geometric constraints must be regarded to make a design at all feasible. Then a complete new - computer aided - recreation of the design so far only conceivable from models or a computer aided recreation of the design adhering now exactly to the constrains imposed by the method of execution, pre manufacturing or the material, becomes imperative.

The methods in detail to be applied vary considerable from case to case and normally no standard solutions exist. In spite of existing highly developed CAD programs, nearly always specific new computer tools and modules have to be created from scratch, tailored to the individual and often unique tasks of the project. This renders this type of work highly challenging and gratifying.

References

1. Linkwitz, K. (1976): Vermessungsarbeiten beim Brückenbau nach dem Taktschiebeverfahren, Proceedings of the VII. International Course on the Application of Geodesy to Engineering, TH Darmstadt.
2. Wenzel, F., Frese, B., Bartel, R. (1987): Die Holzrippenschale in Bad Duerrheim; "Bauen mit Holz", Heft 5, 1987.
3. Gründig, L., Linkwitz, K., Bahndorf, J., Ströbel, D. (1988): Geodätische Methoden als Mittel zur Form- und Zuschnittsberechnung von Flächentragwerken - Beispiel Solebad Bad Dürrheim; Ingenieurvermessung 88, Ferd. Dümmler Verlag Bonn, 1988, ISBN 3-427-78091-0.
4. Linkwitz, K. (1996): Timber Shells - Design, Formfinding, and Premamufacturing; Demonstrated at Three Selected Structures. Proceedings of the INTERNATIONAL WOOD ENGINEERING CONFERENCE, New Orleans Oct 96, Volume 1, Omnipress 2600 Anderson Street Madison, WI, USA.
5. Linkwitz, K. (1996): Holzschalendaecher Hoelderlin-Haus-der-Anthroposophia, Maulbronn; Bautechnik, Heft 4, Ernst & Sohn, Berlin 1996, ISSN 0932-8351.
6. Linkwitz, K. (1998): Execution of a demanding energy saving optimisation measure at the Timber Shell Bad Dürrheim; Proceedings of the 5th World Conference on Timber Engineering, Presse polytechnique et universitaire, Lausanne 1998, ISBN 2-88074-387-7.

CHAPTER 10

CRYSTALLINE ARCHITECTURE

Arthur L. Loeb

Senior Lecturer and Honorary Associate on Visual and Environmental Studies and Member of the Faculty, Graduate School of Education, Harvard University, Cambridge MA 02138, USA

Architects such as Ton Alberts and R. Buckminster Fuller have used the adjective organic to indicate that their designs were based on forms found in nature. Specifically, whereas traditional twentieth-century architecture tended to make use of stacked two-dimensional designs, the new millennium is beginning to derive structures from stacked three-dimensional polyhedra. The right angle, which is rare in natural forms, and unstable, is eschewed by these architects. Crystallographers and solid-state scientists, attempting to understand the reasons why crystalline forms are the way they are, and working in the same Euclidean space on a much smaller scale than architects, tend to have their own perspective on that space. We present here a number of concepts from crystallography and molecular spectroscopy relevant to architecture and vice versa in the hope that the spatial repertoire in both disciplines will be enriched.

1. Introduction

This paper is dedicated to the memory of the late Dutch architect Ton Alberts, who gained special celebrity through his ING Bank headquarters in southeastern Amsterdam, and the Gasunie building in Groningen. In the impersonal wasteland of the Bijlmermeer, Alberts's ING headquarters and the adjacent town square *De Amsterdamse Poort* constitute an oasis of humanity. Characteristic of Alberts's architecture are the undulating and inclined surfaces of his structures, and the use of five-fold rotational symmetry, none of them part of the traditional architectural vernacular.

Alberts called his architecture *organic* because it reflected structure found in nature, and used natural phenomena such as the angle of incidence of sunlight and the storage of energy in plants. His designs thus blend with their contexts, and provide the occupants an environment in which they can be comfortable and adjust an optimal temperature, humidity and illumination.

The American architect and philosopher R. Buckminster Fuller[6,7] was concerned with the conservation of material resources as well as energy, being of the opinion that metal and mineral resources can always be recycled, using the virtually unlimited solar energy entering our global environment. His dome structures and the tensegrity structures invented by Fuller and his student Kenneth Snelson permit dwellings to be mobile, and to shelter large areas from climatic variations.

These architects needed a new vernacular,[28] for designing buildings that met their criteria for functional, aesthetical buildings. Both eschewed the traditional stacked two-dimensional floor plans, and instead worked directly in three dimensions. They took their cue from natural structures[34] which have proven durable and aesthetically pleasing; this paper aims to survey some of the grammar underlying the fundamentals of what has become known as *Design Science.*[30,31,33] Because relevant publications are scattered over a large variety of journals and books, we have appended an extensive bibliography. We shall see that material science has contributed to this grammar, but has in turn benefited from discoveries and inventions made in non-traditional architecture.

2. Symmetry

The world of crystals is very democratic: atoms and ions of a given element are identical to each other, and expect identical contexts.[24,25,34,40,42,44] Symmetry is the quantitative parameter describing the way in which identical components in a pattern are constituted with respect to each other.[2,10,11,26,29]

Translational and *rotational* symmetry relate components that are directly congruent to each other, whereas mirror and glide symmetry relate oppositely congruent components. *Color symmetry*[26,29] relates components that are mutually congruent, but are colored differently. *Dynamic symmetry* will be discussed in the next section.

Two parallel axes of rotational symmetry imply a third axis of rotational symmetry, but the symmetry values of these axes, *k, l* and *m* are limited by the equation

$$1/k + 1/l + 1/m = 1 \tag{1}$$

Accordingly, periodically repeating patterns are limited to the rotational symmetry values 2, 3, 4 and 6, and since the value 5 is not a solution of this equation, there can only be a single axis of five-fold rotational symmetry. Alberts, Fuller as well as the Israeli architect Zwi Hecker[12] have designed structures having five-fold symmetry, but which are not periodic.

Perfect symmetry can be tedious; it represents a state of equilibrium. Slight deviations from symmetry will create tensions toward equilibrium. We tend to perceive more symmetry than is actually present: human faces are generally imagined to be mirror-symmetrical when actually they are not. An interesting example of symmetry breaking occurs in the recently restored royal palace Het Loo near Apeldoorn in the Netherlands. Although the palace and its gardens are mirror-symmetrical, the window in the study of former Queen Wilhelmina does not have a symmetrically situated partner. However, apparently a trained crystallographer was the first visitor, and possibly the only one to date to observe this breach of symmetry.

3. Quasi-Symmetry

About twenty years ago,[3,39,54-56] some alloys were found whose X-ray diffraction patterns exhibited five-fold symmetry. Counter to commonly held beliefs, these patterns did not imply that these alloys themselves were five-fold symmetrical, for diffraction patterns are primarily generated by the interaction between nearest atoms. The mathematician

Roger Penrose had already demonstrated[8] that the Euclidean plane can be tiled by two rhombuses, one having angles 36° and 144°, the other 72° and 108°, such that the symmetry appears to be five-fold locally, but does not extend throughout the pattern. Analogously, three-dimensional space may be filled by two kinds of parallelipipedeons, yielding models for alloys whose symmetry has no five-fold rotational symmetry, but whose X-ray diffraction pattern does. Such crystals are called *quasicrystals*. Linear quasi-symmetrical strings are of interest to the designer.

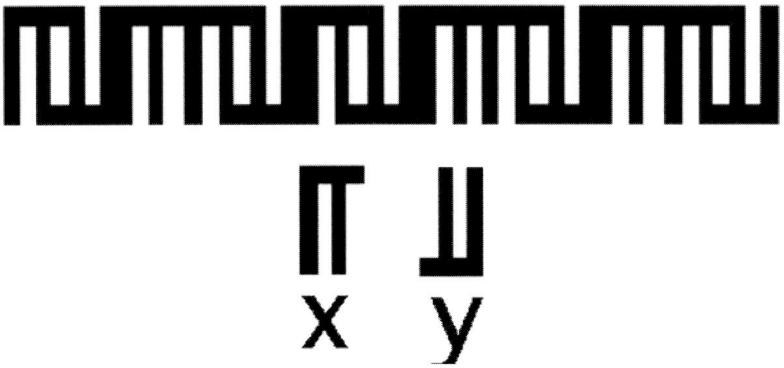

Fig. 1. Quasi symmetrical string.

Take, for instance, the pattern in Figure 1, which has an apparent irregularity near the middle. It was generated by the transformation y->x, x->xy as follows:

y
xy
xyx
xyxxy
xyxxyxyx
xyxxyxyxxyxxy
etc.

Such strings have two notable properties, namely that:

(a) The number of characters, the number of xes, and the number of ys all equal Fibonacci numbers, and
(b) Each string equals the previous one followed by the next previous one.

Although apparently irregular, such a quasi-symmetrical string is actually quite predictable.

4. The Adamantine Angle

When asked about the angles of inclination of Alberts's walls, his collaborator and widow, Lani Van Petten, replied that it was most likely the angle 72° associated with five-fold symmetry. In practice this angle does not differ substantially from the dihedral angle *arc cos (1/3)* of the regular tetrahedron, which is slightly under 71°, but conceptually there is a world of difference. The regular tetrahedron and octahedron together can fill space, whereas the polyhedra having five-fold symmetry, such as the regular dodecahedron, the icosahedron and the rhombic triacontahedron because of their symmetry, do not.[44,48,51] Therefore, multi-storied buildings such as the *Gasunie* and the *ING* would be apt to use the tetrahedral angle.

The difference between these two angles is illustrated in Figures 2, 3 and 4. The first of these figures is a cross section of the regular tetrahedron shown in Figure 3. The triangle ABP is isosceles, and has the following remarkable properties:

(c)

(i) The point O, the center of the tetrahedron, is equidistant from its vertex *P* as from its base *AB*.
(ii) Since the point *Q* is itself the center of a face of the tetrahedron, the length *PQ* equals one third the length PB.

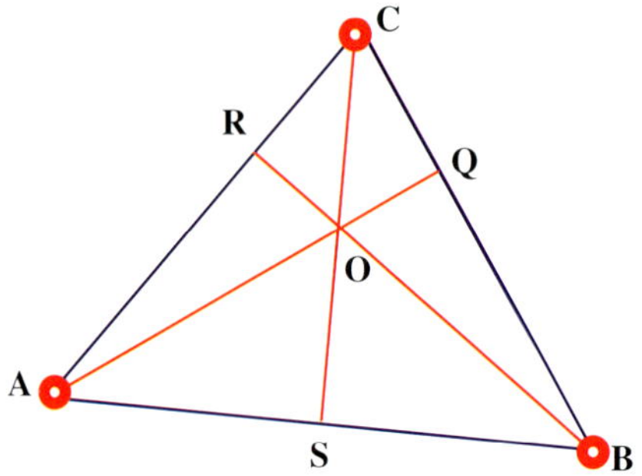

Fig. 2. Mirror plane bisecting the regular tetrahedron.

The cosine of the dihedral angle of the regular tetrahedron, *APB*, therefore equals 1/3. Because this angle is the angle between valence bonds of the carbon angle in organic compounds as well as in diamond, it is of great interest to natural scientists, and is known as the *adamantine angle*.

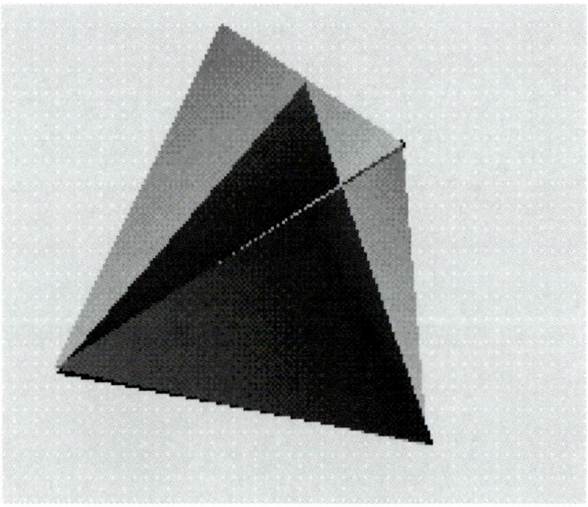

Fig. 3. Bisection of the regular tetrahedron.

The geometry of icosahedron[41,47] is characterized by the golden fraction, ø, which is defined by the equation $1/ø \equiv 1 + ø$.

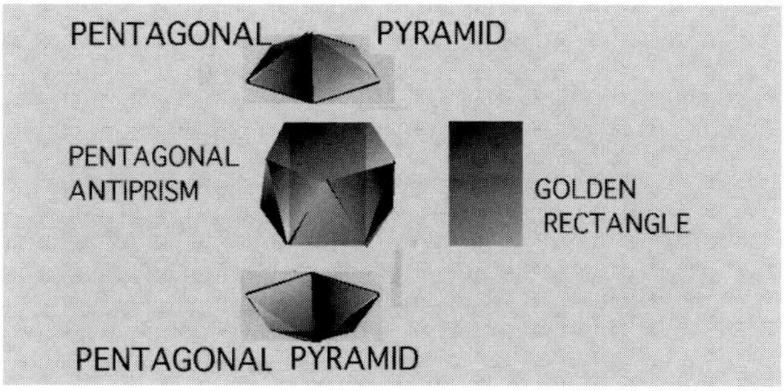

Fig. 4. The regular icosahedron deconstructed.

5. Vector Equilibrium and Stability

In many crystals larger ions tend to surround themselves by twelve identical ions situated at the vertices of a cuboctahedron (Figure 5).[25] Buckminster Fuller called this polyhedron *the vector equilibrium*[7,27] because its edge length equals the distance of each of its vertices from its centre. The cuboctahedron is not a space filler, but together with a regular octahedron does fill space in a 1:1 ratio.

In molecular spectroscopy it is necessary to know the modes of vibration of atoms in a molecule. From this analysis comes the notion of *degrees of freedom*: a polyhedron having V vertices and E edges has Ø internal degrees of freedom,[35,45] where:

$$\text{Ø} = 3V\text{-}E\text{-}6.$$

Such a polyhedron can only be stable if Ø<0. Accordingly, the cuboctahedron, having twelve vertices and twenty four edges, has six degrees of freedom, hence is unstable. Fuller stabilized it by placing a vertex in the center, connecting it to its twelve neighbors, thus in effect creating an octet truss. We have found, however, that in a space frame made up of stacked cuboctahedra having octahedral interstices, the number of edges increases faster with increasing size than three times the number of vertices, with the result that the number of degrees of freedom decreases and the stability of such frames increases with size. Such frames are lighter and more open than octet trusses because there is no vertex in the centers of the cuboctahedra.

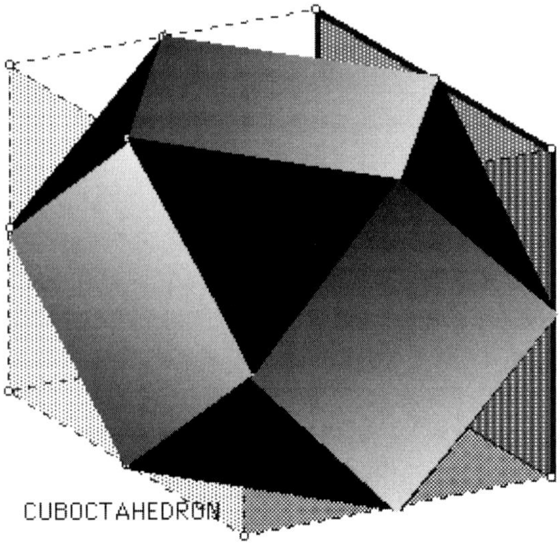

Fig. 5. Cuboctahedron inside a cube.

6. Buckminsterfullerene

In 1985, Robert F. Curl, Harold W. Kroto, and Richard E. Smalley[5] discovered a new form of elemental carbon, having the chemical formula C_{60}. Kroto was working in microwave spectroscopy when he became interested in long chains consisting solely of carbon and nitrogen. At this time, Smalley was studying cluster chemistry. Smalley had designed a cluster beam apparatus which was able to vaporize almost any known material. On September 1st, 1985 the three got together in Smalley's laboratory; in trying to devise the structure of the new carbon molecule, Smalley recalled having bought a Fuller dome kit for his young son, and was able to use this kit to construct a model for the new carbon molecule, which was accordingly named *Buckminsterfullerene*.

It is possible to tile the plane with regular hexagons, but a sphere needs some polygons having fewer sides. When a sphere is tiled by a combination of hexagons and pentagons, regardless of the number of hexagons the number of pentagons must be exactly twelve. A soccer ball has twelve pentagons and twenty hexagons; many domes have more hexagons, as do carbon molecules found subsequently, for example C_{70}.

7. Tensegrity

Buckminster Fuller and Kenneth Snelson invented a class of structures comprising compression members which do not touch, but are held in equilibrium by tension members; each compression member is surrounded by a loop of tension members into which other compression members hook. Particularly stable is the six-strut tensegrity, of which an octant is shown in Figure 6. The other octants are reflections of adjacent octants across the cartesian planes. Three pairs of mutually parallel compression members lie parallel to the cartesian axes; the tension members lie along the edges of the shaded triangle. The distance between parallel compression members equals exactly half of their lengths at equilibrium, and any attempt to alter that distance, *either* by pushing compression members together *or* by pulling them apart, will increase the length of the tension members, which will resist these attempts, holding the structure in equilibrium. When one pair of compression members is made to approach each other, the other ones will also approach each other; when they are pulled apart, the others will again follow suit.

Once more, these architectural inventions have influenced molecular science. Donald Ingber and his associates[4,13-17,57,58] have applied the tensegrity concept to molecular biology with much success.

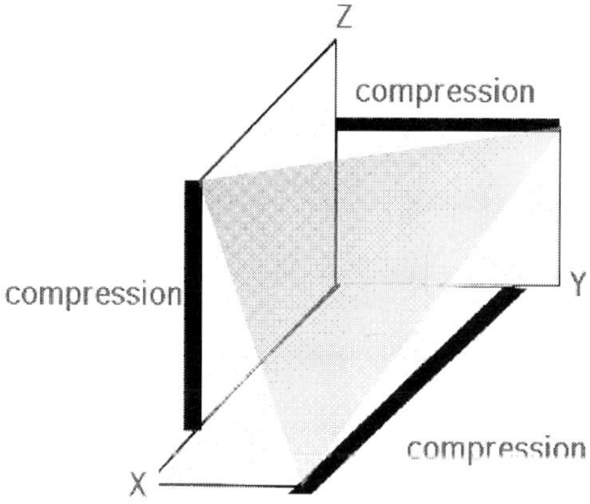

Fig. 6. Octant of a six-strut Tensegrity structure.

8. Conclusions

We note that the fundamental equations governing these examples from the grammar of Design Science are quite simple and elegant. Although upon initial analysis natural structure might appear to be complex, it will be understood if perceived as generated from simple modules by a simple generating rules or algorisms.[36] Alberts, for instance, generated his building designs by a dynamic process, working with the client on a three-dimensional model while the design work was in progress. Our three-dimensional space is not a passive vacuum, but poses constraints. Only when these constraints are well understood, can one build the extensive repertoire permitted by these constraints.

References

1. Adams, M. & Loeb, A.L.: Thermal Conductivity: Expression for the Special Case of Prolate Spheroids. J. Amer. Ceramics Soc. 37, 73-73, 1954.
2. Baglivo, Jenny & Grtaver, Jack A.: Incidence and Symmetry in Design and Architecture, Cambridge University Press, Cambridge, U.K., 1983.
3. Cahn, J. & Taylor, J.: An Introduction to Quasicrystals, Contemp. Math. 64, 265-286, 1987.
4. Chen, C.S. & Ingber, D.E.: Tensegrity and mechanoregulation: from skeleton to cytoskeleton. Osteoarthritis and Articular Cartilage, 1999; 7/1: 81-94.
5. Curl, Robert F., Kroto, Harold W. & Smalley., Richard E. cf the Intrnet under Kroto.
6. Edmondson, Amy: A Fuller Explanation, the Synergetic Geometry of R. Buckminster Fuller, 1986.
7. Fuller, R. Buckminster: Synergetics Macmillan, New York, 1975.
8. Gardner, Martin: Penrose Tiles to Trapdoor Codes, W.H. Freeman, New York, 1988.
9. Goodenough, J.B. & Loeb, A.L.: A Theory of Ionic Ordering, Tetragonal Phase Formation, Magnetic Exchange and Lamellar

Precipitation due to Covalent Forces in Spinels Phys. Rev. 98, 1955.
10. Hargittai, I. & M.: Symmetry through the Eye of a Chemist, VCH Publ., New York, 1987.
11. Hargittai, Istvan, ed.: Symmetry, Unifying Human Understanding, Pergamon, New York 1986 & 1989.
12. Hecker, Zvi: Polyuhedral Architecture, The Israel Musum, 1976.
13. Ingber, D.E.: The Architecture of Life. Scientific American Jan 1998; 278:48-57.
14. Ingber, D.E.: Cellular Tensegrity: defining new rules of biological design that govern the cytoskeleton. J. Cell Sci. 1993; 104:613-627.
15. Ingber, D.E. & Jamieson, J.D.: Cells as tensegrity structures: architectural regulation of histodifferentiation by physical forces tranduced over basement membrane. In: Andersson LC, Gahmberg CG, Ekblom P, eds. Gene Expression During Normal and Malignant Differentiation. Orlando: Academic Press; 1985; 13-32.
16. Ingber, D.E.: Tensegrity: the architectural basis of cellular mechanotransduction. Annu. Rev. Physiol. 1997; 59:575-599.
17. Ingber, D.E.: Biological design principles that guide self-organization, emergence, and hierarchical assembly: from complexity to tensegrity. In: "Unifying Themes in Complex Systems: Proceedings of the 1st International Conference on Complex Systems" Y. Bar-Yam, ed. Perseus Books: New York. 2000; pp. 269-278. [Also available on the web at: Interjournal of Complex Systems 1997; manuscript #146, www.interjournal.org.]
18. Loeb, A.L.: Thermal Conductivity II: Development of a Thermal Conductivity Expression. J. Amer. Ceramics Soc. 37, 1954.
19. Loeb, A.L.: Thermal Conductivity VIII, A Theory of Thermal Conductivity of Porous Materials. J. Amer. Ceramics Soc., 37, 96-99, 1954.
20. Loeb, A.L.: A Binary Algebra Describing Crystal Structures with Closely Packed Ions Acta Crystallographica, 11, 496-476, 1958.
21. Loeb, A.L.: A Modular Algebra for the Description of Crystal Structures. Acta Crystallographica, 15, 219-226, 1962.
22. Loeb, A.L. & G.W. Pearsall: Moduledra Crystal Models. Amer. J. Physics, 31, 190-196, 1963.

23. Loeb, A.L.: The Subdivision of the Hexagonal Net and the Systematic Generation of Crystal Sructures Acta Crystallographica, 17, 179-182, 1964.
24. Loeb, A.L.: The architecture of Crystals, in Gyorgy Kepes, ed.: Module, Proportion, Symmetry, Rhythm, Vision and Value Series, Braziller, New York, 1966.
25. Loeb, A.L.: A Systematic Survey of Cubic Crystal Structures J. Solid State Chemistry, 1, 237-267, 1970.
26. Loeb, A.L.: Color and Symmetry John Wiley & Sons, New York, 1971.
27. Loeb, A.L.: Preface and Concluding Section to R. Buckminster Fuller's Synergetics Macmillan, New York, 1975.
28. Loeb, Arthur L.: Space structures, their Harmony and Counterpoint, 1991 (Originally published Addison Wesley, Reading MA 1976).
29. Loeb, A.L.: Color Symmetry and its Significance for Science, and other contributionsin Patterns of Symmetry, ed. G. Fleck & M. Senechal, U. of Massachusetts Press, 1977.
30. Loeb, A.L.: Algorithms, Structure and Models in Hypergraphics, David Brisson, ed., AAAS Selected Symposium Series, 49-68, 1978.
31. Loeb, A.L.: Sculptural Models, Modular Sculptures in Structures of Matter and Patterns in Science, ed. Marjorie Senechal & George Fleck, Schenkman, Cambridge Mass., 1979.
32. Loeb, A.L.: A Studio for Spatial Order, Proc. Internat'l Conference on Descriptive Geometry and Engineering Graphics, Amer. Soc. Engineering Education, 13-20, 1979.
33. Loeb, A.L.: Structure and Patterns in Science and Art, Leonardo, 4, 339-345, 1971 Reprinted in VisualArt, Mathematics and Computers, ed. Frank Malino, Pergamon, 1979.
34. Loeb, A.L.: Natural structure in the Man-made Environment. The Environmentalist, 2, 43-39, 1982.
35. Loeb, A.L.: Vector Equilibrium Synergy. Space Structures, 1, 99-103, 1985.
36. Loeb, A.L.: Symmetry and Modularity J. Computers and Mathematics with Applications 128, 63-75, 1986. Reprinted in Symmetry, Unifying Human Understanding, ed. I. Hargittai,

Pergamon, New York, 63-75, 1986 (International Series in Modern Applied Mathematics and Computer Science, E.Y. Rodin, Series Editor).

37. Loeb, A.L.: Polyhedra: Surface or Volume? in Shaping Space, ed. M. Senechal & G. Fleck, Birkhäuser Basel/ Berlin/ Boston, 1987. (Design Science Collection, A.L. Loeb, Series Editor)
38. Loeb, A.L.: Review of E.J. Applewhite's Synergetics Dictionary Leonardo, 20, 98-99, 1987.
39. Loeb, A.L.: The Magic of the Pentangle: Dynamic Symmetry from Merlin to Penrose J. Computers and Mathematics with Applications, 17, 33-48, 1989.
40. Loeb, A.L.: Hierarchical Structure and Pattern Recognition in Minerals and Alloys Per. Mineralogia, 59, 197-217, 1990.
41. Loeb, A.L., Gray, J.C. & Mallinson, P.R.: On the Icosahedron, the Pentagonal Dodecahedron and Rhombic Triacontahedron Symmetry, 1, 29-36, 1990.
42. Loeb, A.L.: Reflections on Rotations, Connections and Notations Symmetry: Culture and Science, 1, 39-56, 1990.
43. Loeb, A.L.: Forces shaping the Environment we Inhabit, Internat'l J. Space Structures, 6, 281-286, 1991.
44. Loeb, Arthur L.: Concepts and Images, Visual Mathematics, Birkhäuser, 1992.
45. Loeb, A.L. & Varney, W.: A Stabilized Cuboctahedron Frame Internat'l J. Space Structures, 7, 83-90, 1992.
46. Loeb, A.L.: Geometry and the Imagination in The Visual Mind, 15-17, MIT Press, Cambridge, Mass., 1993.
47. Loeb, A.L.: The Golden Fraction tan [(1/2) arc tan 2], a Tribute to Ernö Lendvai, in the journal Symmetry of the International Society for the Interdisciplinary Study of Symmetry, 5, 303-304 (1994).
48. Loeb, Arthur L.: Buckminster Fuller versus the Irrational, A Double Entendre, in Morphology and Architecture, a special issue of the International Journal of Space Sructures, ed. Haresh Lalvani, 11, 141-154 (1996).
49. Loeb, A.L.: Self-similarity in Quasi-symmetrical Structures, in The Future of Fractals, ed. Clifford Pickover. St. Martin's Press (1996).

50. Loeb, A.L.: A Visual Mathematics Laboratory/ Design Science Studio at Harvard-Radcliffe in Mathematical Sciences Research Hot-Line 1, #4, 1-4 (1997).
51. Loeb, A.L.: Buckminster Fuller and the Relevant Pattern in Beyond the Cube ed. J.F. Gabriel, (John Wiley, 1997).
52. Loeb, A.L.: Deconstruction of the cube in Beyond the Cube ed. J.F. Gabriel, (John Wiley, 1997).
53. Morris, I.L. & Loeb, A.L.: A Binary Algebra Describing Crystal Structures with Closely Packed Ions, Part II Acta Crystallographica, 13, 434-443, 1960.
54. Nelson, David R.: Quasicrystals, Scientific American American, 255, 42, 1986.
55. Senechal, Marjorie & Taylor, Jean: Quasicrystals The Vieuw from les Houches, The Mathematics Intelligewncer, 12, 2, 54-64, 1990.
56. Shectman, D., Blech, I., Gratias, D. & Cahn, J.: Metallic Phase with Long-range Orientational Order and no Translational Symmetry, Phys. Rev. Letters, 53, 1951-1954 (1984).
57. Stamenovic, D., Fredberg, J., Wang, N., Butler, J. & Ingber, D.: A microstructural approach to cytoskeletal mechanics based on tensegrity. J. Theor. Biol. 1996; 181:125-136.
58. Wang, N., Butler, J.P. & Ingber, D.E.: Mechanotransduction across the cell surface and through the cytoskeleton. Science 1993; 260:1124-1127.

CHAPTER 11

FLAT GRIDS DESIGNS EMPLOYING THE SWIVEL DIAPHRAGM

Carolina Rodriguez[1], John Chilton[2] and Robin Wilson[3]

[1]*Lecturer, School of Architecture, University of Liverpool,*
[2]*Professor of Architecture and Design, Nottingham Trent University,*
[3]*Associate Professor, The School of the Built Environment, The University of Nottingham*
[1]*C.Rodriguez-Bernal@liverpool.ac.uk*
[2]*john.chilton@ntu.ac.uk*
[3]*Robin.Wilson@nottingham.ac.uk*

This work is part of the PhD thesis *THE SWIVEL DIAPHRAGM: a geometrical examination of an alternative retractable ring structure in architecture* by C. Rodriguez Bernal undertaken at The School of the Built Environment, The University of Nottingham between 2001-2006.[6] It explores various designs of flat kinetic grids employing the Swivel Diaphragm system. It evaluates the advantages of using this alternative deployable ring structure over other similar systems within grid configurations. The principal aim of this study is to establish a variety of choices for grids offering diverse types of movement. This procedure is unlikely to be exhaustive since the options of grid configurations are so numerous. Hence, in order to avoid a long and repetitive exercise of searching for new configurations, the exploration process has been limited to three different methods. The results of this design process serve as a platform for future research into the possible applications of these types of grids within architectural devices.

Introduction

Scissor structure is a generic name given to certain types of kinetic systems that make use of a particular mechanism where two rigid components are connected by a rotational hinge or pivot and move freely against each other. When various series of scissors are interconnected within more complex assemblies they change in shape and size as a result of the collective action of all the mechanisms working together. Such behaviour is further enhanced when large numbers of units are operating simultaneously. The relationship between the pivots in the mechanism determines the type and direction of movement achieved by the scissor grid. Numerous variations can be made to the basic scissor mechanism in order to modify its kinetic characteristics. Amongst the most widely known are: straight-scissors with a central pivot, straight-scissors with an offset pivot, angulated-scissors and multi-angulated-scissors. There are also different methods for linking scissors within grid configurations. In the particular case of grids employing angulated or/and multi-angulated scissors, there are two main methods that have been developed to configure the grids. The fist involves several concentric layers of scissors that produce a large radial expansion. This method was developed by C. Hoberman *et al.*[1] and was used for his 'Retractable Iris Dome Project'. Other researchers have also studied grids of this type, for example, P. Kassabian, S. Pellegrino and Z. You *et al.*,[3] and F. Jensen.[2] The second method proposes grids with identical deployable rings linked one to the other. As a result the individual rings work in a co-ordinated cyclical sequence of movement where closed and open positions of the grid are not necessarily established. This method was fist explored in the research 'Metamorphic Architecture' by C. Rodriguez.[4] In both methods the grids present difficulties when attached to an external support because the expansion of the grid magnifies or decreases during the deployment process. Hence, it is necessary to use rails or movable supports in order to support the grid.

A recently developed alternative system, named 'Swivel Diaphragm' achieves a similar retraction to that of angulated-scissor and at the same time provides fixed external supports within the ring configuration. Such an attribute helps to overcome support inconvenience associated with the use of angulated-scissors. This system was first proposed by C. Rodriguez and J. Chilton.[5] The present paper explains the ongoing research, exploring the design of flat grids employing this new system.

1. Brief Explanation of the Swivel Diaphragm System

The Swivel Diaphragm [SD] comprised of a concentric series of angulated and straight elements, linked together through elementary pivot joints, as illustrated in Figure 1. Such an assembly forms a closed circuit where all the elements expand and contract simultaneously from or towards the centre of the structure. Therefore, any force applied to a single component is spread throughout the rest of the system. This type of movement is comparable to that of an angulated scissors ring.[1] However, in a SD the components rotate around external fixed supports during the deployment. Thus, the polygonal shape formed by the outer joints will not vary from the closed to the open position. This characteristic is of significance when designing flat grids since the overall structure can be fixed to the support without the need for rails or complex mechanisms.

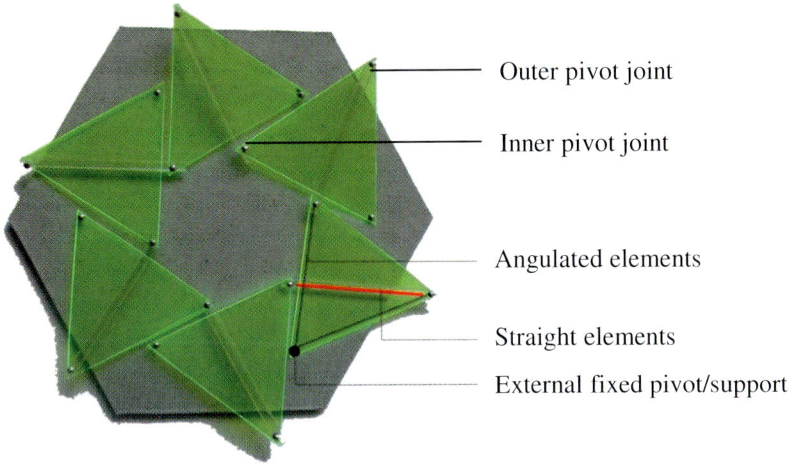

Fig. 1. Hexagonal Swivel Diaphragm with plates.

The number of elements in a SD depends on the desired polygonal shape for the ring. This system admits several regular and irregular polygonal configurations. The present work shows examples of SD and grids using the SD based on basic regular polygon (Figure 2). Other SD based on more complex regular and irregular polygons can be found in Ref. 6. The total deployment of the ring depends on the degree of rotation of each angulated element from its fixed in position support. There are various types of deployments possible with the SD. In the examples of Figure 2 the maximum deployment is achieved when the angulated elements meet or are aligned with the centre of the ring. In these types of rings the deployment is proportional to the number of sides in the polygon. Thus, the more sides in the figure, the less the degree of rotation needed in the angulated elements to achieve a maximum deployment of the ring. This theoretical degree of rotation can be calculated as follows:

$$\text{Maximum degree of rotation} = 360/n \qquad (1)$$

n = number of sides in the polygon

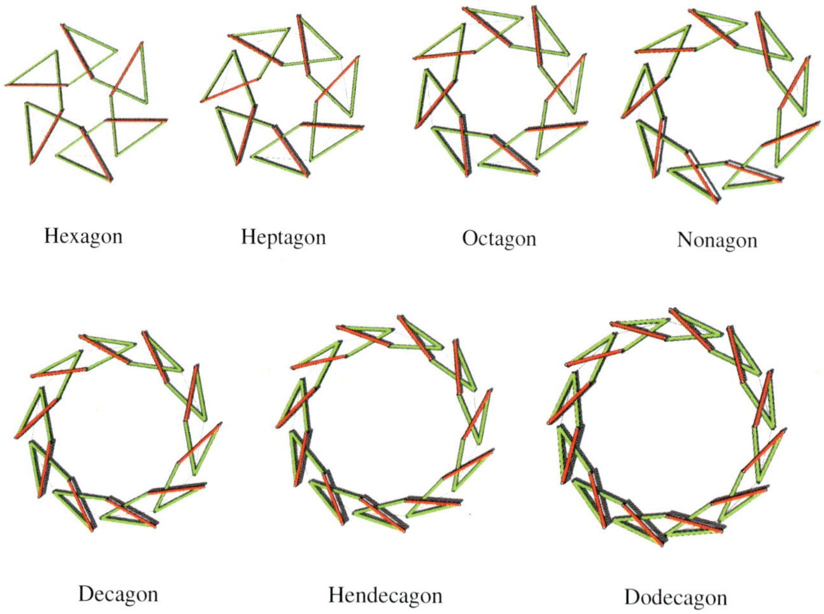

Fig. 2. Examples of polygonal configurations with the SD.

Rings with the SD can adopt three different bar configurations as represented in Figure 3 using distinct linkage patterns but employing an equal number of elements and preserving the same type of deployment and polygonal shape.

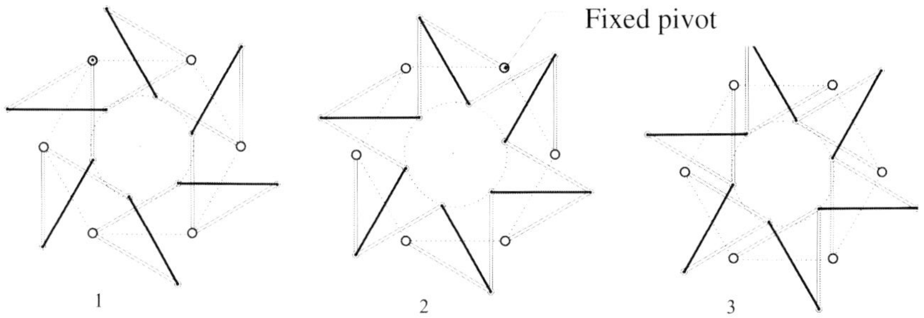

Fig. 3. Possible bar configurations.

2. Flat Grids

2.1. *Diaphragm Morphology*

For physical and practical applications the morphology of the swivel diaphragm may vary depending on the use, the scale, and/or the loading of the structure. The members that constitute the diaphragm can adopt nearly any shape as long as the relationship between the pivots is preserved and overlapping of the elements is avoided. As may be seen in Figure 1, each angulated element compromises three pivots that form a triangle; the fixed pivots transport the loads to the external support, and the other two help to hold up the straight elements that link each module to the proceeding. The load conditions differ when the diaphragm is placed vertically, horizontally or tilted. The diagrams in Figure 4 show the loading distribution in the horizontal and vertical positions. Precise loading and actions in Figure 4 will depend on the detailed structural design of the diaphragms and link bars. Using the appropriate bending moment diagrams as guidelines the members of the SD can be sculpted into more efficient shapes eliminating any unnecessary excess of material. This is of central importance for this type of structure, since substantial weight will significantly affect the retraction of the overall system. It is also ideal to reduce the joint size to the minimum possible in order of reaching the maximum deployment for the diaphragm. Section 2.2 illustrates various 3D proposals developed using different polygonal SDs.

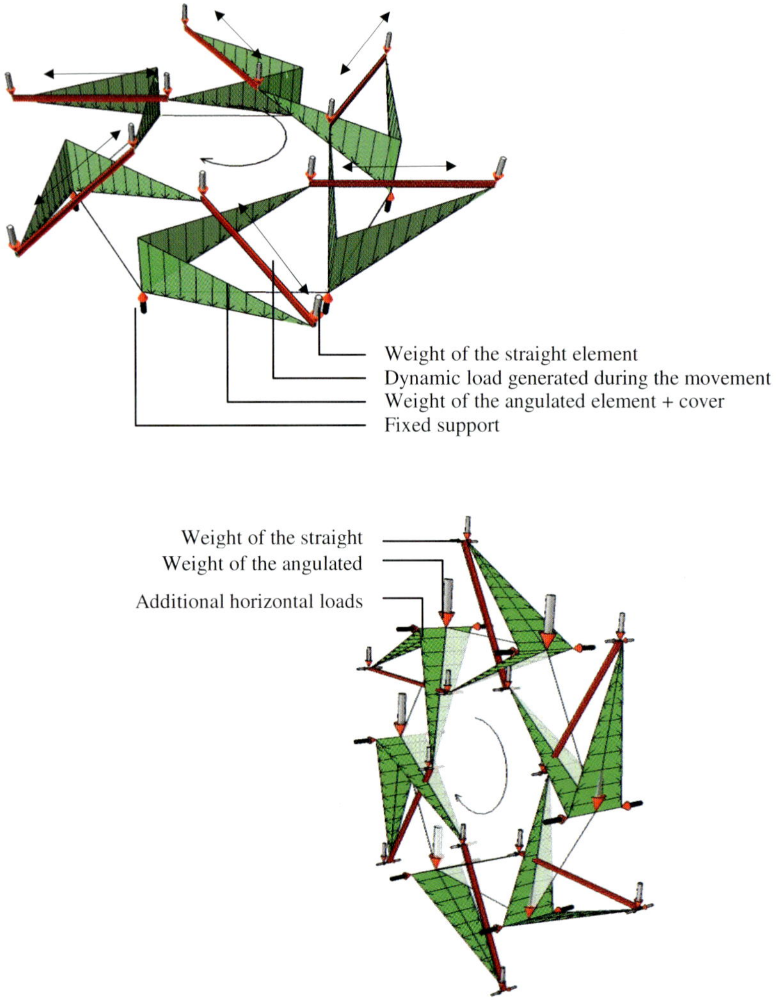

Fig. 4. Diagram of loading in horizontal and vertical position.

2.2. *Module Designs*

Figure 5 shows a hexagonal SD positioned vertically. The depth of the hexagonal frame that supports the ring has been increased to provide more stability to the structure. The diaphragm deploys internally. The angulated elements are designed with a open tetrahedral shape with rigid edges and lightweight surfaces where the maximum height is near the fixed pivot. The linking bars run inside the tetrahedron.

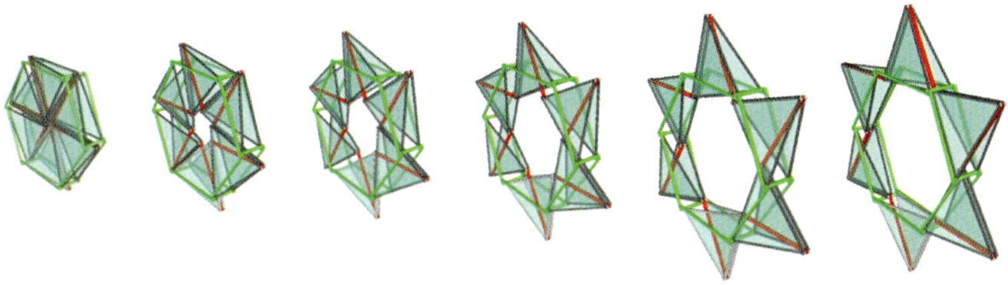

Fig. 5. Vertical hexagonal SD.

Figure 6 shows a heptagonal SD positioned horizontally. The covers are conical segments that meet at the centre of the ring in the fully closed position. The angulated elements compromise sections of arcs that support the conical covers. The arcs are tensed in the base by trusses. Straight bars link the underside of neighbouring elements.

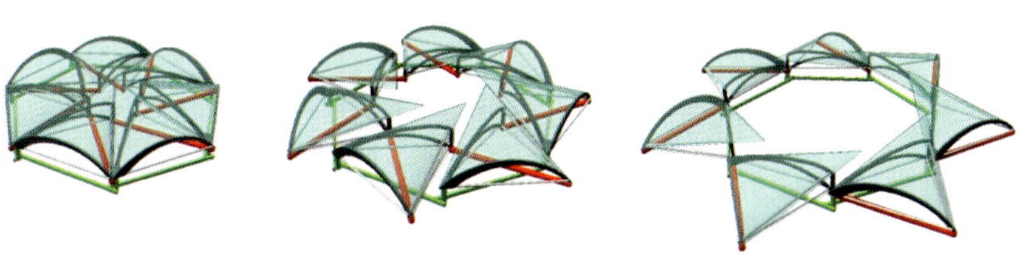

Fig. 6. Heptagonal SD with curved beams and covers.

Figure 7 shows an octagonal SD positioned horizontally. Two triangular beams form the angulated elements and the straight bars that link them run over the top. The design of the plates allows a complete coverage when the ring is fully closed.

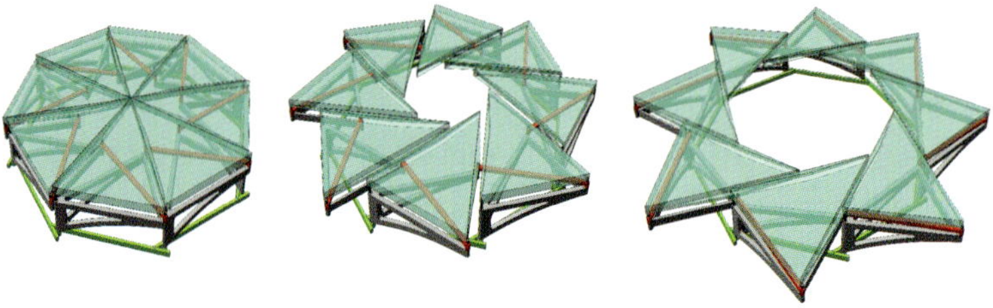

Fig. 7. Octagonal SD with plate covers.

Figure 8 shows a decagonal SD positioned horizontally. The angulated elements are suspended from the fixed pivot position and the covers are tilted forming a turbine shape when the diaphragm is fully closed.

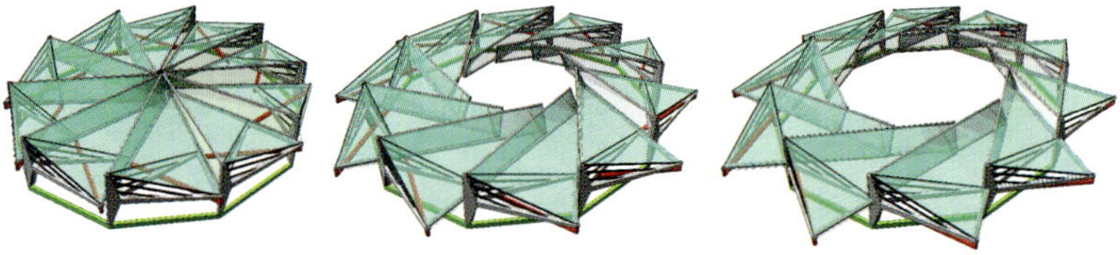

Fig. 8. Decagonal SD with suspended beams and tilted plates.

As demonstrated in the examples of SD modules shown above, some elements of the ring have a different structural function than others. For example, the angulated elements carry the weight of the covers and the straight element plus dynamic loads generated during the process of movement, whilst the straight elements suffer mainly axial compression or tension during the process of transporting forces from one element to the next. Additionally, the pivots need to absorb loads in different directions and at the same time allow rotation. Materials that are lightweight and offer high strength are the most effective for frames of this type, whereas the covers can be built with light and flexible material such as textiles or other synthetic materials.

3. Interconnections for Flat Grids

In flat networks of swivel diaphragms, the modules can be operated either by individually employing an independent mechanism of motion or they can be interconnected to function cooperatively. When the rings are interconnected they may share components such as bars, plates or joints, saving cost in production and assembly. The beauty of interconnected grids lies in the reciprocity of the movement; therefore any force applied to a single component activates the motion of the overall grid. However, in physical models the friction of the pivots and/or the weight of the rigid elements can affect the transmission of these forces from one component to the next. To reduce this problem, the movement could be activated from various points strategically placed throughout the structure.

The possibilities of grid design with SD are limited only by the imagination of the designer. The direction and type of motion can be manipulated depending on the use of the grid or the desired effect. For example, the grid could combine rings moving clockwise with rings operating anti-clockwise or *vice versa* so that when some rings open, others close. Alternatively, the diaphragms can be added concentrically to create radial movement, or, if not, they may be placed more organically to produce disordered kinetic sensations. Despite this variety of choice, it is important that the rings are strategically and carefully placed to avoid interfering with the adjacent rings. Three methods of grouping swivel diaphragms within grids, developed by the authors are illustrated here. The three proposals use the same swivel diaphragm configuration but achieve completely different grid operation. Other methods used to create two and three dimensional grids can be found in Ref. 6.

3.1. *The Fractal Method*

In this method various SDs are jointed concentrically around a central SD. As a result the overall grid twists towards/from a common centre. The central ring is of first order; hence one angulated element is placed on each side of the polygon described by the fixed pivots. The next ring is of second order; therefore two angulated elements are placed on each side of the polygon. The subsequent rings increase in order using the same pattern. With this method the grid behaves as a fractal, the assembly can grow endlessly and at any point of the process the shape of the grid will always be the same as the polygonal figure used for the first central ring. The hexagonal grid shown in Figures 9 and 10 is of third frequency; the angulated elements spin around fixed pivots linked by straight bars. Groups of straight bars operating over the same axis move in unison, thus the bars can be fused into one longer bar connected to various angulated elements. The following equations serve to calculate the number of components needed for the grid according to its polygonal shape and frequency:

Number of angulated elements = $p * 0.5f * (f+1)$ (2)

Number of straight bars = $p * f$ (3)

Number of joints in the straight bar = $f+1$ (4)

f = order of the grid, p = number of sides of the desired polygonal shape

Fig. 9. Hexagonal SD grid generated to third order using the fractal method.

Fig. 10. Model of a hexagonal SD grid generated to third order using the fractal method.

3.2. *The Hexapus Method*

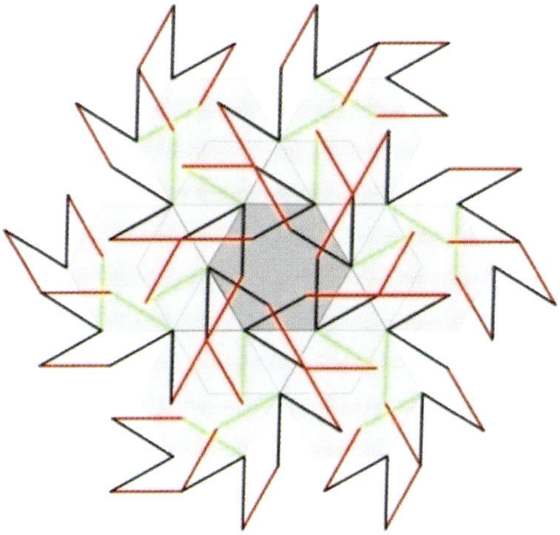

Fig. 11. Hexagonal SD grid using the hexapus method.

Compared with the fractal method, this is a less mathematically based and a more 'organic' technique of linking SDs. It involves creating an arm from each side of the central ring using more complex multi-angulated elements. The central SD is the only completely closed ring, the others are fractions of a ring that form a network of polygons when the grid is fully open and a 'hexapus' shape when the grid is fully closed. Notice that other types of angulated bars have replaced some of the straight bars.

3.3. *Equivalent Link Method*

This is a completely different approach to those described above: in this method isolated SDs are linked by additional straight bars. These additional bars can be placed in many ways as long as they join equivalent pivots in each diaphragm. For example, in Figure 12 pivot 1 is joined pivot 1in the proceeding ring, pivot 2 is jointed with pivot 2 etc. When the grid is fully closed it forms separate polygons with blank spaces in between, whilst when the grid is fully open identical interlinked 'star' shapes are formed.

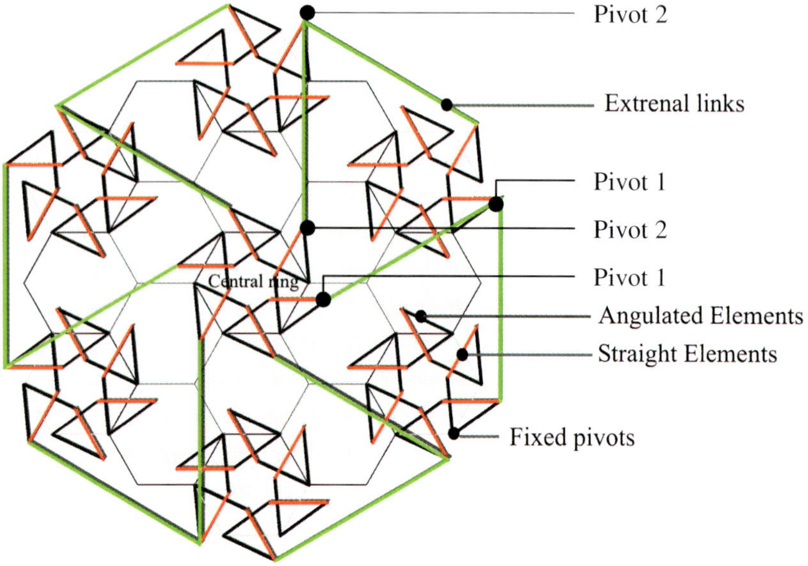

Fig. 12. Hexagonal SD grid using the external link method.

Fig. 13. Equivalent link method grid with hexagonal SD.

Conclusion

The Swivel Diaphragm system shows potential to be used with in flat grid configurations. It offers important structural advantages due to the use of fixed supports. Additionally, it allows very different configurations of grids, providing the designer with a very wide range of options to choose from. Further research needs to be carried out regarding the capabilities of these types of grids with in practical applications. This paper is a step towards achieving this and aims to be a platform to help advance further exploration. The authors strongly believe that there is potential for these grids to be used in architectural devices such as retractable roofs or screen systems for day lighting control. However, this has to be explored through future research.

References

1. Hoberman, C. Web Page, Available At: URL> WWW.Hoberman.Com, Last Accesed: 01/10/03
2. F. V. Jensen, (2001), Cover Elements for Retractable Roof Structures, First Year Report, PhD University of Cambridge.
3. P. Kassabian, Z. You, and S. Pellegrino, (1997), "Retractable Structures Based on Multi-Angulated Elements", in Proceedings of the International Colloquium Structural Morphology: Towards the New Millennium, Chilton J.C et al (eds), University Of Nottingham, UK, 92-99.
4. C. Rodriguez, (2000), Arquitectura Metamórfica, Eds. ICFES-National University of Colombia, Bogotá, Colombia.
5. C. Rodriguez and J. Chilton, (2003), "Swivel Diaphragm a New Alternative for Retractable Ring Structures", in Proceedings of the International Symposium on New Perspectives for Shell and Spatial Structures, Extended abstract, 254–255, and paper on CD-ROM, Taipei, Taiwan.
6. C. Rodriguez, (2006). THE SWIVEL DIAPHRAGM: a geometrical examination of an alternative retractable ring structure in architecture, Thesis submitted to the University of Nottingham, UK for the degree of Doctor of Philosophy, 224p.

CHAPTER 12

FORM-OPTIMIZING IN BIOLOGICAL STRUCTURES — THE MORPHOLOGY OF SEASHELLS

Edgar Stach

University of Tennessee, College of Architecture and Design,
1715 Volunteer Boulevard, Knoxville, TN 37996, USA
Fax 001- 865 9740656
stach@utk.edu

The purpose of this case study is to analyze, the structural properties of natural forms in particular seashells based on digital methods. This is part of a larger architectural study of lightweight structures and form-optimizing processes in nature. The resulting model was used to visually display the internal structure of a seashell and also as input to structural analysis software. The generated renderings show the algorithmic beauty of sea shells.

1. Introduction

Henry Moseley[1] started in 1838 the study of seashells and in particular the mathematical relationship that controls the overall geometry of shells. He was followed by many researchers such as Thompson,[2] Raup,[3,4] Cortie,[5] and Dawkins[6] and others.

The shape of seashells is caused by a logarithmic natural growth (Fig. 4-5). There are three basic shapes: the Planispirally coiled shell (Fig. 1), the Helically coiled shell (Fig. 2), and the Bi-valve shell (Fig. 3). Environmental factors such as the availability of raw materials, the type of substrate, the amount of calcium present, as well as many other factors contribute to deviations from the three basic forms.

Helically coiled gastropod shell

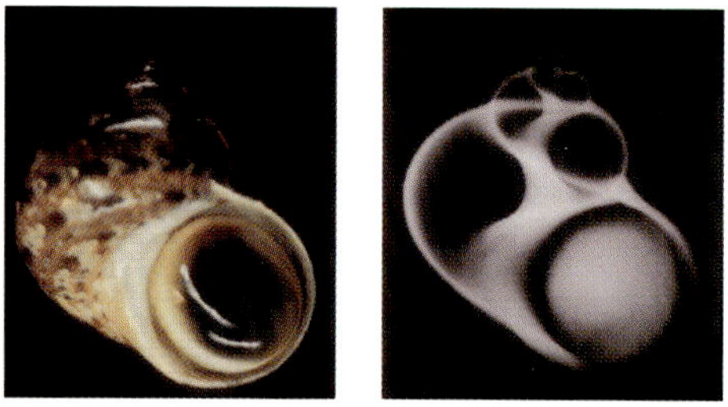

Fig. 17. Elevation and X-ray views of helically coiled gastropod shell (Conklin p. 153).

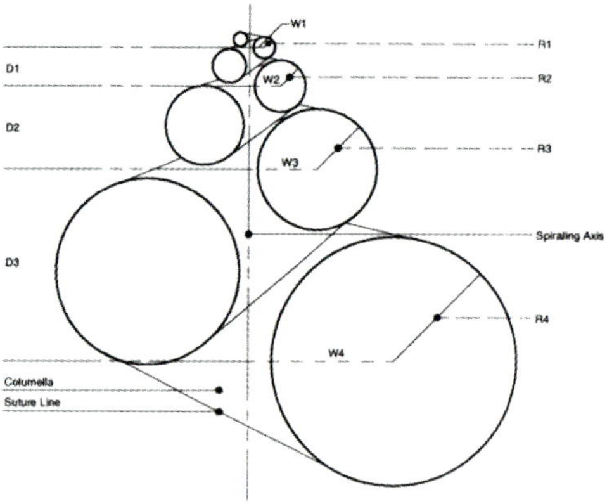

Fig. 18. Diagram of proportioning system of helically coiled gastropod shells.

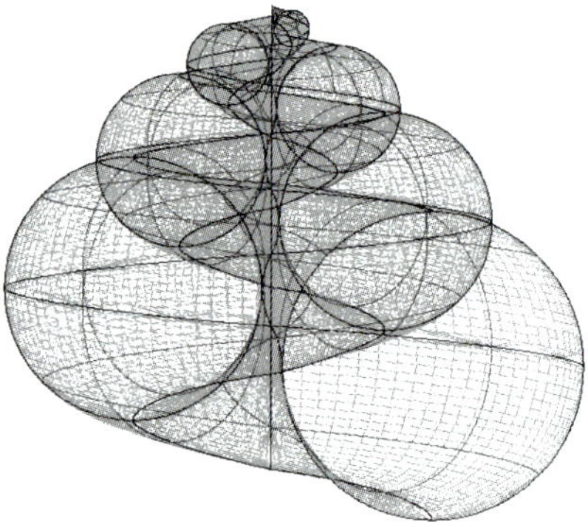

Fig. 19. Computer wire frame model of helically coiled gastropod shells.

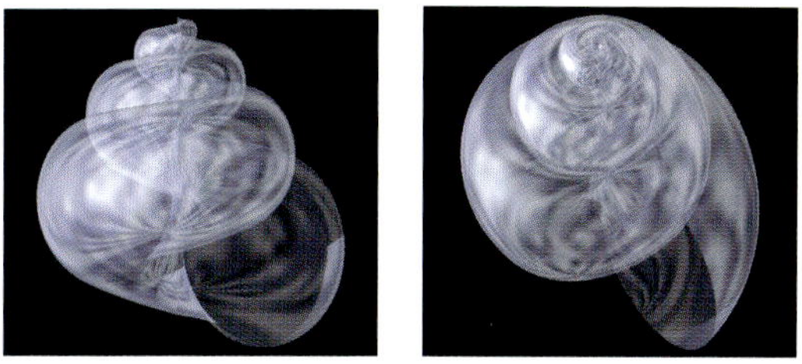

Fig. 20. Rendered computer model of helically coiled gastropod shells.

Bi-valve shell

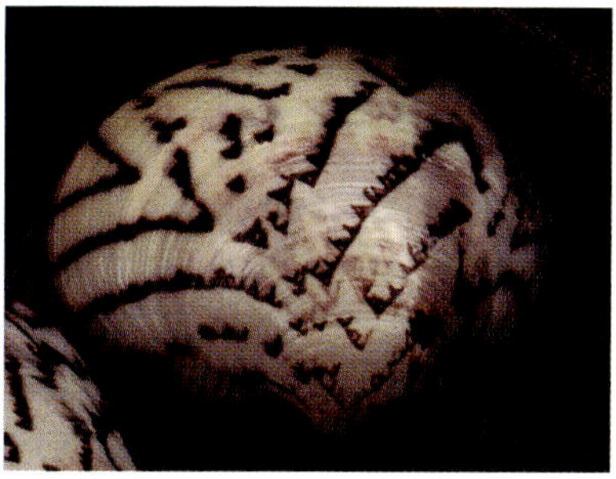

Fig. 21. View of Bi-Valve Shell (Harasewych p. 209).

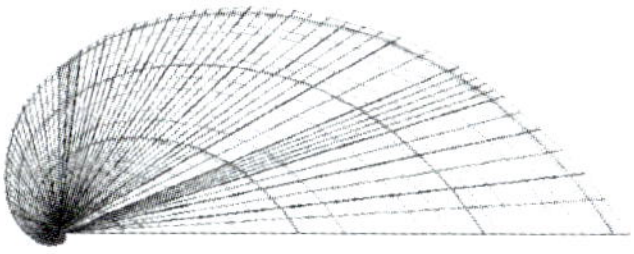

Fig. 22. Elevation of computer wire frame model showing exponential curve of shell growth.

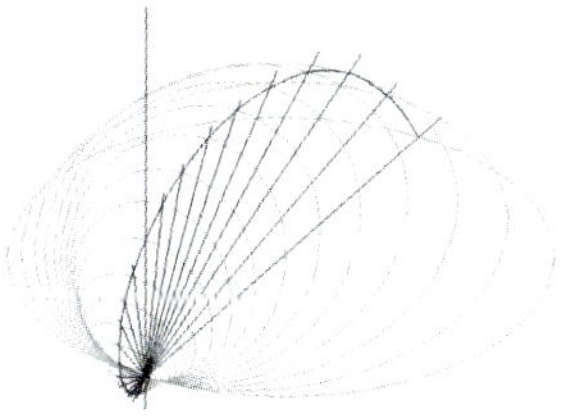

Fig. 23. Three dimensional representation of proportioning system used to develop curvature bi-valve shell.

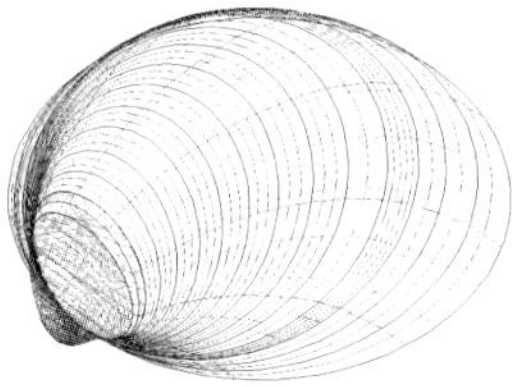

Fig. 24. Computer wire frame model of of bi-valve shell.

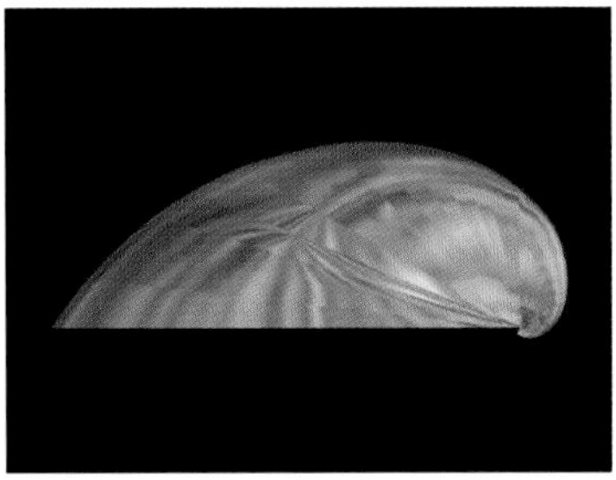

Fig. 25. Rendered computer model of bi-valve shell.

6. Structural Optimization in Engineering — Seashell Structural Properties

The shell geometry responses to any external and internal loads by redirecting forces within a very thin section of shell structure along its natural multiple curvatures. Finally these forces are transferred to the supported area such as ground, rock or sand depending upon how the seashell positions itself in the environment.

Genetic algorithms

In engineering fields, accomplishing an objective with a minimum of effort, either in terms of material, time or other expense, is a basic activity. For this reason it is easy to understand the interest designers have in different optimization techniques like the seashell structure. Mathematical, as well as, model based tools have traditionally been employed for such optimization. In recent times, mathematical methods executed on computers have become predominant. Unfortunately, computer derived solutions often obscure the range of possible solutions from the designer by only exhibiting a final, 'best' solution. Naturally, optimization methods can only respond to the objective parameters which are coded into the problem, and as a result, non-coded parameters, such as aesthetics, or context are left out of the optimization process, and ultimately left out of the final design solution.

Structural optimization in engineering takes natural constructions as an example. Similar to nature itself, computer-generated genetic algorithms can be calculated using stated goals to achieve global optimization - the search strategy is, like in nature, goal-oriented. An evolutionary algorithm maintains a population of structures (usually randomly generated initially), that evolves according to rules of selection, recombination, mutation and survival, referred to as genetic operators. A shared 'environment' determines the fitness or performance of each individual in the population. The fittest individuals are more likely to be selected for reproduction (retention or duplication), while recombination and mutation modifies those individuals, yielding potentially superior ones. Using algorithms, mechanical selection, mutation and recombination improves generationally with a fixed parameter size and quality.

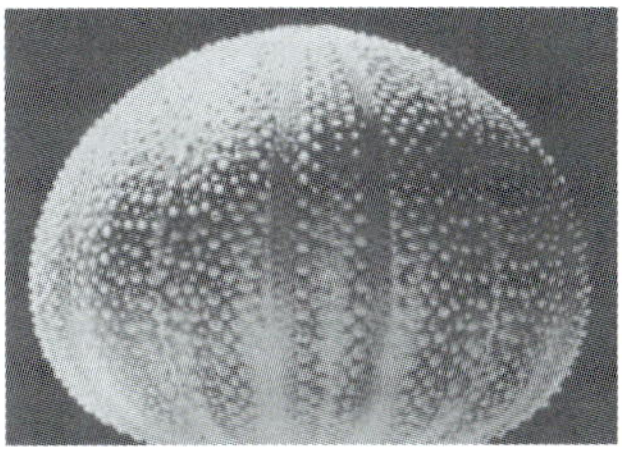

Fig. 26. Structure optimization in the shell structure of a sea urchin.

Fig. 27. Finite Element analysis of sea urchin shell, color coded stress analysis [Process und Form, K. Teichmann].

Computer-compressed evolution

Design space and finite elements

Computer-compressed evolution like the SKO method (Soft Kill Option) (Fig. 28) follows the same construction principle that nature employs to promote for example the shell growth of a sea urchin (Fig. 26/27) or the silica structure of sea shell (Fig. 1/3). Building material can be removed wherever there are no stresses, but additional material must be used where the stresses are greater. This is the simple principle that evolution has used for millions of years to produces weight optimized 'components'. Using computer programs based on computer-generated genetic algorithms like the SKO method, scientists are now able to simulate this evolution and compress it into a short time span.[9]

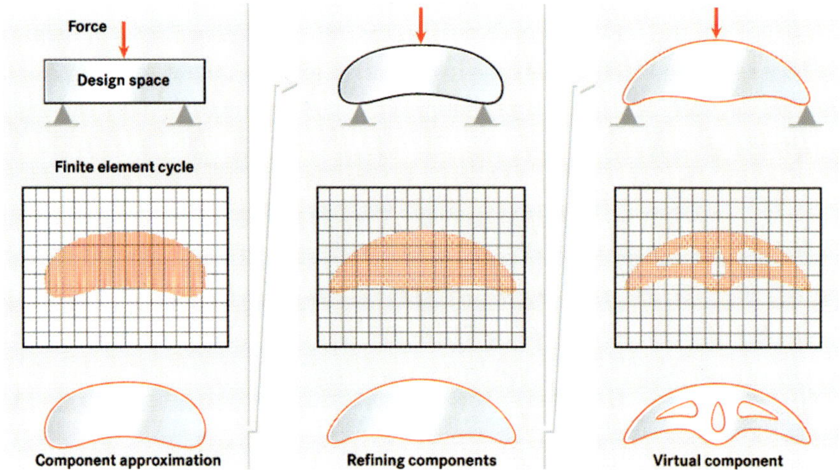

Fig. 28. SKO method (Soft Kill Option). [Hightech Report 1/2003, pp. 60–63].

In order to simulate lightweight engineering strategy according to nature's guidelines, scientists using the SKO method must first define a virtual design space, which represents the outermost parameters of the component being developed (Fig. 27). To subdivide this design space into many small individual parts, the finite elements, a grid is applied. If now a virtually external load applied, the computer calculates the resulting force exerted on every one of the finite elements. The FE model shows exactly where there is no load stress on a component and in turn shows where it is possible to make savings with regard to the materials used. On the other hand, for areas that bear heavy stress the simulation program indicates the need to reinforce the construction material. Like nature the computer repeats this 'finite element cycle' several times. As a result, they can refine a component repeatedly until the optimal form — one that evenly distributes the stresses within a component — is found.

Conclusion

The abstract geometrical properties of seashells can be described by there mathematical relationship. The translation of abstracted nature in mathematical terms and by applying prerequisite architectural considerations is the fundamental concept of form and structure analyses. The value of this research is to develop mathematically definable models of structure systems in nature. The goal is to define a set of structural principles, and to make those principles applicable for architects and engineers.

Seashell structures are perfect study models for self- organization structures in nature because of there relatively simple physical and morphological principle and geometry based on four basic mathematical parameters. Self-organization it the defining principle of nature. It defines things as simple as a raindrop or as complex as living cell - simply a result of physical laws or directives that are implicit in the material itself. It is a process by which atoms, molecules, molecular structures and constructive elements create ordered and functional entities.

Engineers are using this concept already successful for optimization processes in a white range of applications starting in mechanical-, medicine-, air and space engineering. Architects are only one step away adopting the same technique for designing in a macro scale buildings and structures. Material scientists are already designing and producing new materials or smart materials in a Micro scale using the self organizing principles. In the future, the material engineers will develop constructions out of self-structuring materials that consciously use the principles of self-organization, creating not only materials with brand new properties but also inspiring architects to define their constructions in a more intelligent way.

At its best, intelligent structures and materials will influence the entire philosophy of construction. Engineers will no longer ensure safety through quantity of material and cost. Simple structural analysis will no longer suffice; instead, self-organizing structures will define the new construction principles.

Acknowledgements

Research Assistant: John Falko, University of Tennessee, College of Architecture and Design University of Tennessee, College of Architecture and Design Research Grant

References

1. D'Arcy Wentworth Thompson, On Growth and Form, Dover Publications, Inc., 1992.
2. David M. Raup, "The Geometry of Coiling in Gastropods", Proceedings of the National Academy of Sciences of the United States of America, Volume 47, 1961.
3. David M. Raup, "Computer as Aid in Describing Form in Gastropod Shells", Science, July-September 1962.
4. M. B. Cortie, "Models for Mollusc Shell Shape, South African", Journal of Science, 1989.
5. Chris Illert, Foundations of Theoretical Conchology, Hadronic Press, Inc., 1992.
6. Richard Dawkins, Climbing Mount Improbable, W. W. Norton & Company, 1997.
7. Hans Meinhardt, The Algorithmic Beauty of Sea Shells, Springer, 1998.
8. K. Teichmann, Prozess und Form, Ernst und Sohn.
9. HIGHTECH REPORT 1/2003, 60-63.

10. Claassen, Cheryl, Shells, Cambridge University Press, 1998, Cambridge, UK.
11. Conklin, William A., Foreword by Alan H. Shoemaker, Nature's Art: The Inner and Outer Dimensions of the Shell, University of Carolina Press, 1985, Orangeburg, South Carolina.
12. Harasewych, M. G., Photography by Murray Alcosser, Preface by Stephen Jay Gould, Shells Jewels.
13. From The Sea, Rizzoli International Publications, Inc., 1989, New York, New York.
14. Wye, Kenneth R., The Encyclopedia of Shells, Quarto Publishing plc, 1991, New York, New York.
15. Jirapong, Kamon, Krawczyk, Robert J., Case Study: Modeling the Interior Structure of Seashells, ISAMA/Bridges Conference, July, 2003.

CHAPTER 13

EXPANDABLE 'BLOB' STRUCTURES

Franck Jensen[1] and S. Pellegrino[2]

Department of Engineering, University of Cambridge, Trumpington Street, Cambridge CB2 1PZ, United Kingdom

[1]*frank@cantab.net*

[2]*pellegrino@eng.cam.ac.uk*

This paper presents a methodology for designing single-degree-of-freedom expandable "freeform" structures composed of rigid blocks connected through simple cylindrical joints. The underlying idea is to interconnect two or more individually expandable plate structures. Using a two-dimensionally expanding sphere as a first example, the conditions that must be satisfied to preserve the internal mobility when connecting identical expandable plate structures are explained. These conditions are then extended to plate structures that are not identical and it is then shown that a wide range of expandable free-form or "blob" structures can be designed through this approach.

1. Introduction

This paper is concerned with the geometric design of expandable structures consisting of rigid elements connected through cylindrical hinges or scissor joints that only allow rotation about a single axis. The authors have been interested in developing stacked assemblies formed by rigidly interconnecting the expandable plate structures that they had previously developed.[4] The connections between individual plates can themselves be volume filling, and so the stacked structure can also become an expandable three-dimensional object. As the plate structures from which one starts can have any plan shape, and only simple kinematic constraints have to be satisfied in order for them to maintain their internal mobility in the stack configuration, nearly any shape can be generated, including so-called free-forms or blobs.

The approach presented in this paper starts by developing a method for rigidly connecting two identical and individually expandable plate structures, such that the assembled structure only possesses a single-degree-of-freedom. The kinematic constraints that must be satisfied by the connections and the connected plate structures are derived, and are shown to allow the stacking of non-identical plate structures as well, thereby allowing free-form profiles to be obtained for the assembled structure.

This paper is presented as follows. The following section briefly reviews the kinematic properties of the expandable plate structures that are to be connected, and their underlying bar structures. Then, the section An Expandable Sphere describes a method for connecting the plate structures such that they can form an expandable sphere. The next section is then concerned with the kinematic constraints that have been satisfied by these connections and the formulation of a general set of rules for such connections. These are then used in the following section to design an expandable free-form or blob structure demonstrating the possibility of designing vivid and exciting expandable structures using this method. A brief discussion concludes the paper.

2. Background

Simple expandable structures based on the concept of pantographic elements, i.e. straight bars connected through scissor hinges have been known for a long time. One of the simplest forms of such pantographic structures is the well-known lazy-tong in which a series of pantographic elements are connected through scissor hinges at their ends to form two-dimensional linearly extendible structures.

More sophisticated expandable or deployable structures have been developed over the last half century.[1,2,12] Many of these solutions are based on the so-called angulated pantographic element.[3] In its simplest form, shown in Figure 1, it consists of two identical angulated elements each composed of two bars rigidly connected with a kink angle α. Unlike pantographic elements composed from collinear bars, which have $\alpha = 0$, such angulated elements can be used to form expandable closed loop structures if the conditions $\overline{AE} = \overline{CE}$ and $\alpha = 2\arctan(\overline{EF}/\overline{AF})$ are met.[6,10] These conditions guarantee that the following equation of geometric compatibility is satisfied for all deployment angles, γ

$$\tan(\alpha/2) = \frac{\overline{CE} - \overline{AE}}{\overline{AC}}\tan(\gamma/2) + 2\frac{\overline{EF}}{\overline{AC}} \quad (1)$$

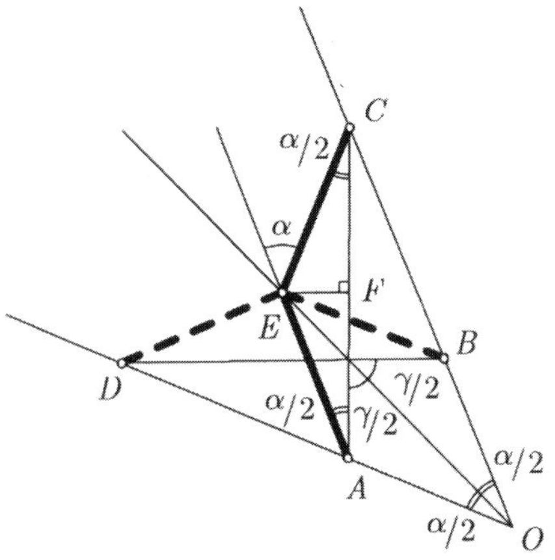

Fig. 1. Pantographic element consisting of two angulated elements, each formed by two bars.

As the kink angle α is constant, i.e. independent of γ, the end joints of the pantographic element move along radial lines, and so an expandable closed loop structure can be formed from these elements.

You and Pellegrino[10] found that certain *multi-angulated elements*, i.e. pantographic elements with multiple kinks, have the same property. Consider the circular structure shown in Figure 2. It consists of two separate layers, shown with solid and dashed lines, each containing 12 identical multi-angulated elements, which here have three kinks. As the structure expands, from left to right in the figure, the hinges and hence both layers of elements expand radially. However, while moving radially each multi-angulated element in the solid-line layer also rotates *clockwise* as can be seen in the figure. The dashed-line layer, on the other hand rotates *counter-clockwise*. Note how the multi-angulated elements form three concentric rings of rhombus-shaped four-bar linkages all of which are sheared as the structure expands.

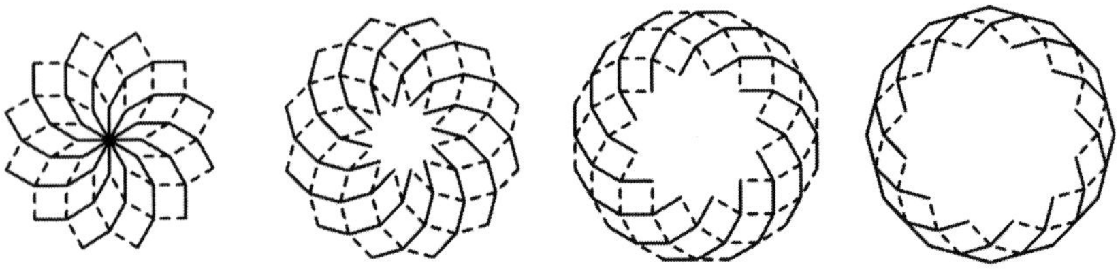

Fig. 2. Retractable structure formed from multi-angulated elements.

Expandable structures composed from rigid plates and scissor or spherical joints have been proposed.[8,9] Kassabian *et al.*[5] proposed covering the above basic circular bar structure with rigid panels to form a continuous, gap free covering. As an extension of his earlier work on bar structures You proposed to replace parts of the bar structure with rigid elements[11]; this idea was further developed by Rodriguez and Chilton in their Swivel Diaphragm.[7]

The present authors have proposed a method for covering any multi-angulated bar structure with plates.[4] By considering the shearing deformation of any four-bar linkage formed by the elements common to two pairs of consecutive angulated elements, the authors determined a general condition on the shape of the boundary between two covering elements, such that the plates would not restrict the motion of the structure while resulting in a gap and overlap free surface in either extreme position of the structure. The covering of a single rhombus is shown in Figure 3. The boundary angle θ for the plate elements is determined by:

$$\theta = \frac{\pi - \beta_{\text{closed}} - \beta_{\text{open}}}{2}. \tag{2}$$

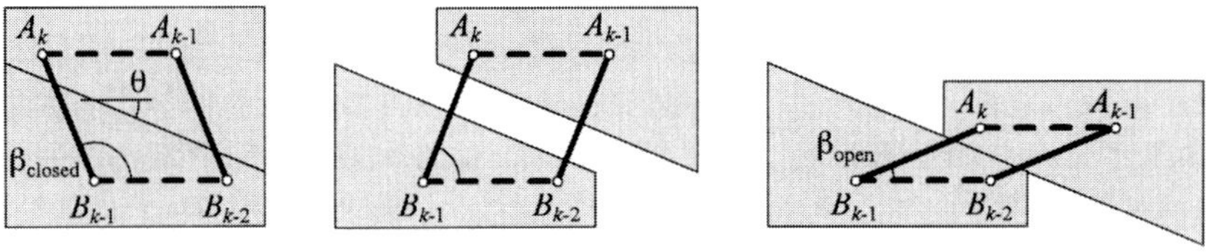

Fig. 3. Movement of four-bar linkage with two plates attached.

In fact, instead of covering the bar structure with plates, it is possible to remove the angulated elements and connect the plates directly, by means of scissor hinges at exactly the same locations as in the original bar structure. Thus, the kinematic behaviour of the expandable structure remains unchanged. It was shown that, as long the plate boundaries have a certain periodic shape, they need not be straight.[4] Figure 4 shows an example of such an expandable plate structure; this non-circular structure is formed by 26 plates of different shape.

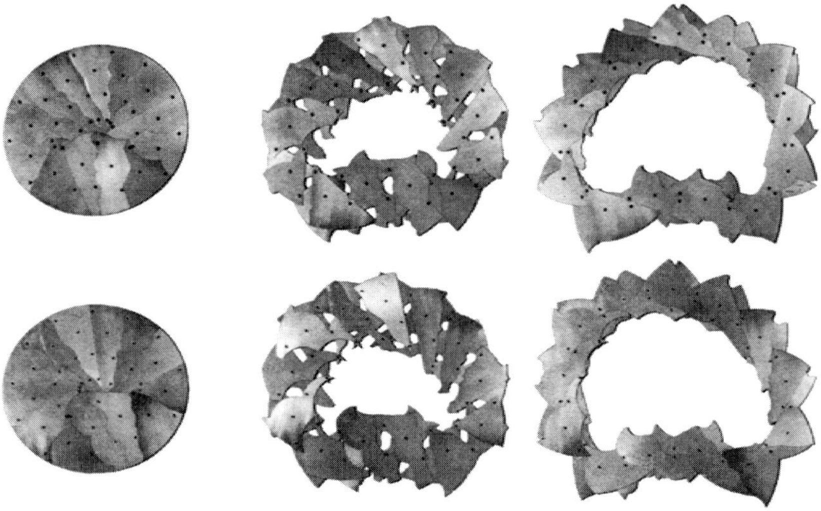

Fig. 4. Model of non-circular structure where all plate boundaries are different; both layers are shown.

3. An Expandable Sphere

To investigate the possibility of creating three-dimensional expandable structures by stacking identical plate structures, a simple design for the plate structures was chosen. The design, shown in Figure 5, consists of 16 identical plate elements of which 8 form the bottom layer and 8 the top layer. As for the bar structure shown in Figure 2, the 8 plates forming the top layer move radially outwards while rotating *clockwise*, while the plates forming the bottom layer (of which only small parts can be seen in the figure) rotate *counter-clockwise*. From symmetry it can be concluded that all plates in the same layer rotate by identical amounts; the rotations of plates in different layers are equal and opposite.

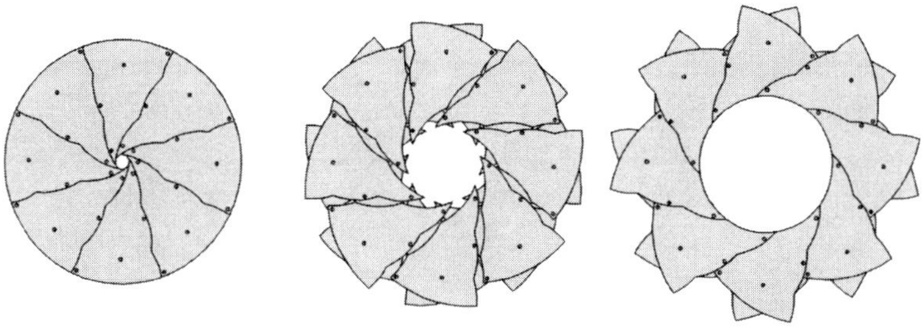

Fig. 5. Expandable circular plate structure.

Consider two such identical plate structures positioned above one another, which are to be rigidly connected. In order to preserve the internal mobility of the two plate structures, the rigid connections must be made from a single plate in one structure to a single plate in the other structure. However, noting the opposite rotations of the two layers in each structure, it is not possible to connect the plates in the top layer of the bottom structure to the plates in the bottom layer of the top structure while maintaining the internal mobility of these structures. Instead, the connections must be either between the top or the bottom layers, as these have identical motions. This can be achieved by swapping the top and bottom layers in one of the two structures so that connections can then be made between the facing and now identical layers of plates. Hence, by rearranging the order of the layers in one of the two structures it becomes possible to rigidly interconnect them, so they form a stacked and still expandable assembly.

A stacked assembly composed of identical plate structures would form a cylinder, but other profiles can be formed by varying the plan shape of the outer edge of the plate structure.

An expandable assembly that approximates to a spherical profile has been designed and built from 5 interconnected plate structures. From the model shown in Figure 6 it can be seen that the 5 structures conform to a spherical profile in the closed position. Note that the periodic boundaries of the plate elements on the upper face of each plate structure are convex in alternate directions, which confirms that the layers have been swapped in alternate structures. All five plate structures have been produced using the same plate template, only the outer boundaries of the plates have been made with different radii. Clearly, the allowable rotation between the extreme positions of neighbouring plates is identical throughout the structure, giving identical motions for the 5 plate structures.

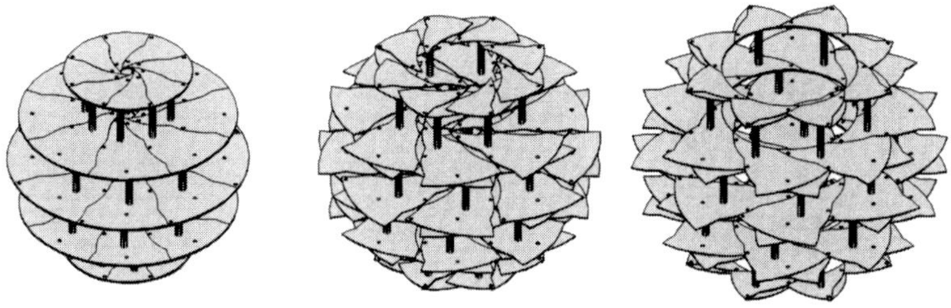

Fig. 6. Perspective views of expandable spherical structure.

In Figure 6 the connections between the 5 structures have been shown as solid rods rigidly attached to the plate elements. However, as the motion of two connected plates is identical, the connectors themselves could be formed in a number of different ways. For example, to form an object with a smooth, continuous surface in the closed configuration, one can connect the plates using either solid or hollow blocks. By curving the outer wall of each block, and also the top/bottom surface of the blocks attached to the uppermost/lowermost plate structure, one can obtain a continuous spherical profile. The internal faces of the blocks could be made perpendicular to the original plate, and follow the periodic edge shape of the plates.

Figure 7 shows such a spherical model. It was constructed using identical plastic plate structures of which four were trimmed so that their outer boundaries form circles of different radii. The connections were made from identical blocks of light foam board, cut using abrasive water-jet cutting. The blocks were then glued to the back of the individual plastic plates that form the plate structures and the spherical profile obtained by removing the excess foam board material.

Fig. 7. Expandable sphere.

4. Stack Structures

To rigidly connect two expandable plate structures the motion of the individual plates being connected must be identical. Earlier, the motion of each plate was described as the combination of a radial motion, i.e. a translation and a rotation. Kassabian *et al.*[5] found that, as the rotation in the two layers is equal and opposite, imposing an additional rigid body rotation to the whole structure, equal to the rotation undergone by one of the layers, the motions of the two layers become a pure rotation and a pure translation, respectively. If, for example, the imposed rotation is such that the plates in the top layer of the bottom structure undergo a pure rotation, then each plate in this layer rotates about its own fixed centre.[5]

Hence, consider two plate structures that are to be connected. Following the above approach impose the same rigid body rotations on both structures; clearly the plates in the two connected layers must *rotate about the same axes of rotation and by the same amounts*. Since the rotations of two rigidly connected plates are automatically synchronised, the only kinematic constraint that must be satisfied is that the two plates have the same axes of rotation. This is the only condition that must be satisfied for any two plates belonging to different expandable plate structures to be connected without suppressing the internal mobility of the plate structures.

It was mentioned in Section 2 that an expandable plate structure is kinematically equivalent to a bar structure with identically positioned hinges. It is therefore possible to identify the centres of rotation for the equivalent bar structure, instead of the plate structures themselves. It is therefore proposed to start the design of an expandable stack assembly by considering its underlying bar structures; a particular embodiment as a plate structure will be determined later on.

Following Hoberman,[3] an expandable bar structure of general shape is generated in its open configuration by considering an n-sided polygon. Each angulated element is then defined so that its central scissor hinge coincides with a vertex of the polygon and its end points coincide with the midpoint of the adjacent polygon sides. Thus, in Figure 8(a) we have defined two noncircular structures; structure I is drawn with a thinner line, structure II is drawn with a thicker line. Structures consisting of multi-angulated elements can be generated by adding any number of rhombuses to this base structure.[10]

Kassabian *et al.*[5] showed that the axis of rotation for any particular angulated element is at the half-way point between the origin of the polygon, *O*, and the vertex defining the central hinge of the element. Hence, all the axes of rotation are at the vertices of a polygon that is half the size of that defining the elements themselves, as shown in Figure 8(a). Therefore, together with the location of the chosen origin, the initial polygon determines the location of the axes of rotation. Or, alternatively, a polygon defining the axes of rotation together with a chosen origin point define a bar structure, and hence also a plate structure that will rotate about those specific axes when it is expanded.

In Figure 8 note that the bar structures I and II have been defined such that three centres of rotation — A_{cen}, B_{cen}, and C_{cen} — are common to the two structures. Hence, the elements A, B and C of the two structures, drawn as solid lines, can be rigidly connected to each other as

International Association for Shell and Spatial Structures, Vol. 44, No. 143, pp 181–188.

8. H.F. Verheyen, A Single Die-Cut Element for Transformable Structures, International Journal of Space Structures, Vol. 8, No. 1&2, pp 127–134.
9. K. Wohlhart, Double-chain mechanisms, In: S. Pellegrino and S.D. Guest, eds., IUTAM-IASS Symposium on Deployable Structures: Theory and Applications, Kluwer Academic Publishers, Dordrecht.
10. Z. You and S. Pellegrino (1997), Foldable bar structures, International Journal of Solids and Structures, 34(15), 1825–1847.
11. Z. You (2000), A new approach to design of retractable roofs, In: S. Pellegrino and S. Guest, eds., IUTAM-IASS Symposium on Deployable Structures: Theory and Applications, Kluwer Academic Publishers, Dordrecht, pp 477–483.
12. T.R. Zeigler (1981), Collapsible self-supporting structures and panels and hub therefore, USA Patent No. 4,290,244.